A PHILOSOPHY FOR THE SCIENCE OF WELL-BEING

A PHILOSOPHY FOR
THE SCIENCE OF WELL-BEING

Anna Alexandrova

OXFORD
UNIVERSITY PRESS

OXFORD
UNIVERSITY PRESS

Oxford University Press is a department of the University of Oxford. It furthers
the University's objective of excellence in research, scholarship, and education
by publishing worldwide. Oxford is a registered trade mark of Oxford University
Press in the UK and certain other countries.

Published in the United States of America by Oxford University Press
198 Madison Avenue, New York, NY 10016, United States of America.

© Oxford University Press 2017

Library of Congress Cataloging-in-Publication Data
Names: Alexandrova, Anna, 1977– author.
Title: A philosophy for the science of well-being / by Anna Alexandrova.
Description: New York : Oxford University Press, 2017.
Identifiers: LCCN 2016048538 | ISBN 9780199300518 (hardcover : alk. paper)
Subjects: LCSH: Well-being. | Happiness. | Quality of life.
Classification: LCC BF575.H27 A44 2017 | DDC 158—dc23
LC record available at https://lccn.loc.gov/2016048538

1 3 5 7 9 8 6 4 2
Printed by Sheridan Books, Inc., United States of America

CONTENTS

CONTENTS

ACKNOWLEDGEMENTS

This book began its life in St. Louis, Missouri, and was completed in Cambridge, England. So it has inherited philosophical trademarks of both cities.

In St Louis I learned much of what I know about the philosophy and the science of well-being from my colleagues at the University of Missouri St. Louis, Washington University in St. Louis, and St. Louis University. Dan Haybron, whose pioneering work on this topic and whose generosity are truly unsurpassed, is my greatest influence. Other knowledgeable guides have been Simine Vazire, John Doris, Adam Shriver, Matt Cashen, Eric Brown, Eric Wiland, John Brunero, Ramesh Raghavan, Philip Robins, and Randy Larsen. When I realised that I cannot bring myself to advocate any particular theory of well-being as most philosophers do, I started searching for ways to justify my indecision, and this is how I stumbled on contextualism. I learned it from Brit Brogaard, Irem Kurtsal Steen, Mark Steen, Gillian Russell, John Greco, and Jim Stone. Gualtiero Piccinini, Carl Craver, Kent Staley, Tyler Paytas, Andrew Black, José Luis Bermúdez, Scott Berman, Jim Bohman, and Ron Munson have given valuable advice that stuck.

By the time I arrived to Cambridge in 2011 I saw this book mostly as a call for philosophy to 'fix' the science of well-being. But this theory-sceptical city changed my plans. Inspired by Hasok Chang, Huw Price, Eran Tal, and Felicia Huppert, I started thinking about measurement and the limits of philosophy. As theory gives out, politics take over, and this has both plusses and minuses as I learned from Steve John, Simon Schaffer, Jim Secord, Tim Lewens, Helen Curry, Tiago Mata, and Lorna Finlayson. During the five years here I have had the tremendous fortune of having the ears of many brilliant and generous people, but especially memorable are comments by Adrian Boutel, Chris Clarke, Arif Ahmed, Jane Heal, Richard Holton, Rae Langton, Caspar Hare, Clare Chambers, Markus Anderljung, Tom Dougherty, Bernardo Zacka, Paul Sagar, Lucy Campbell, Marta Halina, Bence Nanay, Jonathan Birch, Paulina Sliwa, Matt Drage, Gabriele Badano, Trenholme Junghans, Liz Chatterjee, Chris Newfield, Mike Kelly, and the much missed great John Forrester. Elina Vessonen, Willem van der Deijl, Sam Wren-Lewis, and Jacob Stegenga have kindly read the entire draft with great care and insight.

The third city that needs acknowledging is San Diego, California. As a graduate student at the University of California San Diego I flourished due to the perfect combination of challenge and protection, and only now do I realise what hard work this was for my teachers Craig Callender, Naomi Oreskes, Gila Sher, Bob Westman, Steve Shapin, Bill Bechtel, Wayne Martin, Don Rutherford, Branislav Slantchev, Sam Rickless, Paul Churchland, and Jonathan Cohen. I am immensely proud to call Nancy Cartwright my academic mother and my friend. She gave me many gifts but most formative among them are getting to know Sir Stuart Hampshire, not worrying about whether my work is 'core' philosophy, maintaining a default suspicion of unified theories and a default respect for local knowledge, and pages of timely and wise hand-written comments that covered my drafts. Sometime around 2004 she gently suggested that I look into happiness research, an area that had nothing to do with my dissertation. I did, and that is when this project was born. But I would not have made it far if it weren't for a custom reading list put together for me by David Brink and for critical but encouraging comments on my initial work on happiness by Dick Arneson, Dana Nelkin, and Rick Grush.

Elsewhere many people gave me advice, ideas, and support that shaped this work. Among them are Valerie Tiberius, Eleonora Montuschi, Sophia Efstathiou, Alex Michalos, Dan Hausman, Erik Angner, Mike Bishop,

Mary Morgan, Heather Douglas, Elizabeth Anderson, Julian Reiss, Alex Voorhoeve, James Wilson, Anthony Skelton, Dale Dorsey, Matt Brown, Leah McClimans, Rich Lucas, Matt Adler, Dan Benjamin, Ori Heffetz, Matt Adler, Michael Weber, Michael Tooley, Chris Heathwood, Neera Badhwar, Ben Bradley, Ben Bramble, Scott Sturgeon, Stephen Campbell, Francesco Guala, Nellie Wieland, Nina Brewer-Davis, Anca Gheaus, Don Ross, Harold Kincaid, Joshua Knobe, Edouard Machery, Ron Mallon, and John Abraham. I would like to thank Peter Railton in particular for a wise comment that toned down my variantism and Antti Kauppinen for pushing me to test contextualism properly and for the Joey Tribiani example. (Both happened at Sam Wren-Lewis's excellent 2012 workshop in Leeds.) I benefited from many anonymous referees but especially from those my editor Peter Ohlin found for this press and from the remarkably patient and generous reader of my *Philosophy Compass* article, whose exemplar of refereeing I will now try to emulate.

And now for the rocks on which I stand. My husband Robby Northcott has had my back since 1998. Together we concocted a fateful grant application to the Open Society Foundation, which made possible my otherwise unlikely studies in the United States. For 18 years he has been my sounding board, my source of calm and of perspective, my coconspirator, my coauthor, and more recently my beloved coparent of the glorious Andrei and Misha. It is hard to imagine this book without the three of them and my devoted English in-laws. It is even harder to imagine it without my parents Elena and Andrei and my grandparents and great-grandparents. During the perestroika-themed conversations in our tiny Soviet kitchens, they treated all things intellectual with reverence and awe. This is how, even though they spent their days at building sites, factories, and breadlines, I came to love ideas. They made great sacrifices to help me leave my hometown of Krasnodar in the 1990s to study philosophy of all things, but they made these sacrifices with gusto, excitement, and full trust in me. More recently my parents and my sister Dasha enabled me to write this book by providing months of childcare. To this whole crew is my greatest debt.

ORIGINS OF THE CHAPTERS

The Introduction uses some material from 'Well-Being and Philosophy of Science' *Philosophy Compass* 10.3 (2015), 219–231; 'Well-Being' in *Philosophy of Social Science: A New Introduction*, eds. Nancy Cartwright and Eleonora Montuschi (Oxford: Oxford University Press, 2014), 9–30; and 'Well-Being as an Object of Science' *Philosophy of Science*, 79.5 (2012), 678–689. Chapter 1 is based on 'Doing Well in the Circumstances,' *Journal of Moral Philosophy*, 10.3 (2013), 310–328. Chapter 2 started as 'Values and the Science of Well-Being: A Recipe for Mixing' in *The Oxford Handbook of Philosophy of Social Science*, ed. Harold Kincaid (New York: Oxford University Press), 625–645. Chapter 3 is a descendant of 'Toward a Theory of Child Well-Being' *Social Indicators Research*, 121.3 (2015), 887–902, coauthored with Ramesh Raghavan. Chapter 4 is based on 'Can The Science of Well-Being Be Objective?' *British Journal of Philosophy of Science* (2016 axw027. doi: 10.1093/bjps/axw027). Chapters 5 and 6 use material from 'Is Well-Being Measurable After All?' *Public Health Ethics* (2016 phw015. doi: 10.1093/phe/phw015) and 'Is Construct Validation Valid?' coauthored with Daniel Haybron *Philosophy of Science* 83.5 (2016), 1098–1109.

INTRODUCTION

Fitting Science to Values
and Values to Science

Denying the importance of philosophy to science is just as wrong as insisting on its constant and unavoidable relevance. The first extreme fails because science, today an enterprise separated from philosophy, nevertheless makes philosophical bets in every step of the way: concept formation, method choice, confirmation procedures, and so on. The second extreme—and this is a less familiar point—amounts to a failure to learn from history of science. 'Getting over' a philosophical debate, leaving it unsolved and moving on, has been crucial to the production of knowledge at many junctures. In this book I set sail between these two rocks. I want to show which philosophy is indispensable and which can be safely ignored—not in general but only for one important corner of today's social and medical science: *the science of well-being.*

I use this expression as an umbrella term for all research whose goal, at least in part, is a systematic and empirical study of well-being. Typically it goes under the names of quality of life or happiness studies; positive or hedonic psychology; or studies of subjective well-being, life satisfaction, flourishing, and welfare. Some of these scientists would be happy to stand under this umbrella, and to this extent the category I propose already exists and reflects the way many conceive of their work. Indeed *science of well-being* is the name of a 2003 Royal Society

Discussion Meeting whose goal was to encourage this research, already flourishing in North America, in the United Kingdom.[1]

But I also mean the umbrella to cover projects that are *implicitly* about well-being even if scientists themselves do not use this term. For example, when medical researchers study the so-called patient-reported outcomes (PROs), or when economists study material welfare, they sometimes distance themselves from the term 'well-being'. They may do so because patient-reported effectiveness of treatment or consumption are allegedly narrower, less demanding, and more tractable states than well-being. But sometimes this separation from well-being is a poorly veiled attempt to weasel out of the hard questions. Effectiveness to what end? Consumption for the sake of what? It is hard to see how PROs or consumption can be defined without some reference to well-being. For these indicators to be valuable they must bear the right relation to well-being, even if they do not capture the whole of it. So even when well-being is not the direct object of study, it is still a value in which the studies of many other concepts bottom out. Definitions of health often refer to well-being to pick out particular areas of human functioning; norms of rationality acquire their status as normative in part because they suit human pursuits, of which the pursuit of well-being is surely one; economic growth, sustainability, resilience, human capital all have the shape that they do in part because they are supposed to bear on well-being. Indeed the deepest challenges across social science and political and moral theory are often implicitly about well-being. How to organise public science? How to relieve suffering? And, of course, how to live?

Thus my term 'science of well-being' is sometimes the actors' category and sometimes not. But it is the category I postulate because these projects, as we shall see, raise very similar questions. It may sometimes be difficult to say precisely whether a given project is or is not part of the science of well-being. My strategy in this book is to concentrate on rather obvious cases—when well-being or something very close to it is treated as an object of empirical knowledge—but it is likely that what I say about these obvious cases would also hold for less obvious ones.

1. This meeting resulted in an eponymous influential Huppert et al. (2005) volume.

So this book is about well-being as an object of science: how science should define well-being, how it should measure it, and the role of philosophy in all this. Philosophers of science, along with historians and sociologists of science, study how knowledge is and should be produced, whether we can trust it, and how we come to do so. As we shall see shortly, there is now a fully fledged science of well-being. A philosophy of this science is an account of how it is possible, and where and why this knowledge succeeds and fails.

But a philosophy of this particular science needs to be special in one respect. Sometimes a scientific concept has a *value* element in its content: it not only describes but also judges and guides. That's when science and philosophy are entangled in a further way than usual: not just metaphysics and epistemology enter but moral philosophy too. One cannot classify a policy or an outcome as well-being enhancing by merely stating empirical facts or reporting opinions. For any standard or method of measurement of well-being is already a claim about the appropriateness of an action or state in the light of some assumed value.

These features of my project—to comment on a science, but a science laden with judgements about good life—lead me to seek out ears of two audiences. My first conceit is to address the scientists of well-being and those who use this science: as a philosopher of this science I can speak to how to study well-being better and what users should and should not expect from this knowledge. As part of this goal I show that definitions and measures of well-being require substantive and often controversial assumptions that are sometimes hidden behind apparently neutral and technical facts or avoided altogether, all in the name of preserving objectivity. This is wrong epistemically and morally. The science of well-being is better off when its values are well-articulated and defended, as I show possible.

My other conceit is to speak to my fellow philosophers. It is no good clamouring for a greater attention to philosophy if philosophy does not have much to offer. In my view moral philosophy today—a major academic project that proclaims to be studying well-being—could be offering a lot more than it actually does. Philosophers of well-being spend more resources than appropriate chiselling out theories of well-being immune to counterexamples and at too abstract of a level.

That is an exercise that science can safely ignore. Instead progress will come from a different kind of work—contextual theorising about what well-being amounts to in different circumstances that individuals and communities face.

So this book is a proposal for reform in both directions: the science of well-being should never pretend to do without philosophy and philosophy should get its act together and provide usable tools for science. The rest of this introduction gives an overview of the science in question and previews the arguments I make in later chapters.

THE PAST AND THE PRESENT

A history of this science is yet to be written. Although I do not offer one, it is fair to start by acknowledging that well-being has long featured in scientific projects, sometimes as a background motivation, sometimes as an object of knowledge. Today's enthusiasts paint the science of well-being as radically new, path-breaking, or revolutionary. Its creation myths usually represent the scientists of the past as not caring about well-being or not having the proper tools to study it, while today we have both tools and the good sense to do so.[2] Without discounting this pioneering spirit, we should nevertheless not overestimate the novelty of the enterprise.

For starters, concern with human well-being is at the very root of modern social science. The earliest mentions of the phrase *science sociale* in revolutionary France take place in the context of justifying and furthering the ideals of justice and democracy. In 1798, Jean-Jacques-Regis Cambacérès, a statesman and the author of the Napoleonic Code, in his 'Discours sur la science sociale' explicitly identified social science with the means of securing happiness (*bonheur*) for all (Sonenscher, 2009). Social science thus began its life as a form of knowledge devoted officially to the advancement of well-being. Though the precise conceptions of social science differed, its founders in the Enlightenment

2. Some examples are in Seligman (2004, Chapter 2), Seligman and Csikszentmihalyi (2000), Kahneman and Krueger (2006), and Frey (2008).

and nineteenth-century France, Germany, Scotland, and England— Jeremy Bentham, Adam Smith, Nicolas de Condorcet, James and John Stuart Mill, Auguste Comte, Karl Marx—all conceived of social science as central in the project of bringing about happiness, relieving suffering, liberating, furthering progress. And so they shaped the subject matter and the methodologies of the new sciences in part to serve this goal. Psychology would help us measure and predict changes in happiness, sociology to advance society to the next more perfect stage of development, political economy to document how we live and to predict the macro-consequences of the individual actions, be they in pursuit of happiness or not.

In the twentieth century behaviourist concerns with unobservability of mental states purged the language of happiness from social science. Or so the traditional story goes. But the story does not show that well-being fell off the agenda. In economics, happiness was replaced with 'welfare' measured apsychologically but nevertheless measured and studied by means of analyses of consumption and efficiency. A concern with subjective experience is not particularly new either. Attempts to conceptualise and measure subjective well-being were live from about the 1920s in the applied fields surrounding psychology, such as marital and education sciences, and in the social indicators movement of the 1970s (Angner, 2011b). Outside the quantitative tradition, humanistic psychology as practiced by Carl Rogers, Abraham Maslow, and the therapists inspired by psychoanalysis took flourishing, happiness, and self-actualisation as central to their thinking and their work. Finally, the central place of well-being in the medical sciences, nowadays evident in the proliferation of PROs, is reflected in the 1946 World Health Organization's definition of health as 'a state of complete physical, mental and social wellbeing and not merely the absence of disease or infirmity'.[3]

Most recently well-being entered the agenda of social sciences with the discussions of the so called Easterlin Paradox. Formulated by American economist Richard Easterlin in the 1970s (Easterlin, 1974), the paradox juxtaposes two facts. The first fact is that at any given time

3. World Health Organization (1948).

and within any country income predicts self-reported happiness. The second fact is that over time as income increases happiness does not correspondingly do so. Easterlin hypothesised that beyond a certain minimum, people judge their happiness by their relative rather than their absolute income, and this idea spurred a great deal of research on the relationship between objective circumstances and life evaluation, satiation points beyond which money makes no difference, as well as on the psychology of happiness judgements. For several decades the Easterlin Paradox served the role of justifying the policy relevance of the sciences of well-being—after all, if happiness stalls as income grows, focusing on economic growth to the exclusion of other goods seems wrong. Many articles, books, and grant applications to study well-being started by citing Easterlin's landmark study. This equilibrium is now somewhat shaken, as the new data brought out forcefully by economists Betsey Stevenson and Justin Wolfers (2008) appear to undermine the second fact—increase in absolute income does after all predict increase in subjective well-being over time. If income is a fine long-term predictor of happiness, what policy role is there for indicators of subjective well-being? The enthusiasts are undaunted for several reasons. First, Stevenson and Wolfers rely on indicators focused on satisfaction with life relative to other possible lives, rather than on measures of emotional well-being. The latter does not track income as well. Second, income only predicts subjective well-being in conjunction with other social factors such as health, social support, freedom, and so on. The apparent demise of the Easterlin Paradox is unlikely to undermine policy excitement around well-being research. Third, even if on average absolute income and subjective well-being rise and fall together, there are still striking cases of divergence, for example, the steady growth of gross domestic product (GDP) coupled with a steady fall in life satisfaction in Egypt and Tunisia during the Arab Spring.[4]

So a history of this enterprise will be a history of the involvement of scientific knowledge in the projects of social and political improvement

---

4. Easterlin (1974) is the original study; Stevenson and Wolfers (2008) is the critique denying the paradox; Clark at al. (2012) presents the state of the art; Organisation for Economic Co-operation and Development ([OECD], 2013) and Adler and Seligman (2016) defend continued relevance of subjective well-being.

and governance. It will also be a story of expansion of measurement and of quantification of phenomena that were previously thought to be private, idiosyncratic, unmeasurable. These are the themes of historiographies of recent social and psychological sciences (Porter, 1995; Rose, 1990, 1998), and they are readily visible in today's widespread institutionalisation of this science.

This institutionalisation is hard to overstate. Well-being science now boasts of professional societies, specialised journals, research institutes, and publications in prime venues such as *Science*.[5] There is also the sheer quantity: 'well-being' and its cognates regularly top the lists of keywords in scientific abstracts[6], and 'well-being' alone brings up over 5 million entries on PubMed, which is twice as many as 'cancer'.

THE NORMAL SCIENCE OF WELL-BEING

Going along with the institutional there is an intellectual maturity. Today's science of well-being has fairly settled goals and methodologies and increasingly settled empirical facts. These goals, methodologies, and facts are regularly publicised in reviews of latest findings.[7] Thomas Kuhn's (1962) notion of 'normal science' naturally suggests itself. For Kuhn normal science started when fundamental philosophical disagreements ended and paradigm-based puzzle solving began. I do not wish to debate whether the science of well-being has a paradigm in a sense that is defensible or fitting to Kuhn's intentions. But I nevertheless introduce this field by enumerating its commonly shared commitments and in this sense I speak of a normal science of well-being. This way of

5. For societies see the International Society for Quality of Life Studies, International Positive Psychology Association. For journals see the *Journal of Happiness Studies, International Journal of Wellbeing, Applied Research in Quality of Life, Applied Psychology: Health and Well-Being, Social Indicators Research, Journal of Positive Psychology*. For high-profile publications see Layard (2010), Kahneman et al. (2004a).
6. Well-being was the second most popular keyword in all psychology articles cited in the Social Science Citation Index and the Science Citation Index between 1998 and 2005 (Zack & Maley, 2007).
7. US psychologist Ed Diener is undoubtedly the most prolific writer of such field-defining review articles. The latest are Diener (2012), Diener et al. (2016).

introducing the object of my discussion—by focusing on its intellectual rather than material activities, and on the shared, rather than the controversial ones—is not uniquely right, I am happy to admit, but it fits my purpose, as will become clear shortly.

But before I can start, in addition to the idea of a normal science, I need also an idea of a social science. The science of well-being is pursued by sociologists, economists, psychologists, anthropologists, medical, legal, business and social work scholars—that is, mostly social scientists. Now philosophies of social sciences have traditionally fallen into two camps. The first camp advocated a kind of *exceptionalism*. *Interpretivists*, the exceptionalists par excellence, insisted that social science has a distinct goal of understanding human action by the method of interpretation, which may not allow for a great deal of generalisable knowledge. The second camp—*naturalists*—emphasised the continuity of social with the natural sciences, emphasising the search for laws and causal explanations.[8] Recently, philosophers and historians of science noted that natural sciences are too diverse to have a monolithic method. Indeed, the many observations of the *disunity of science* that grew through the 1980s and 1990s should have already doomed this way of carving up naturalism from interpretivism. Perhaps the real debate, as Daniel Steel (2010) claims, is about whether or not generalisable causal knowledge can be attained and used for the betterment of human lives, with interpretivists arguing that it cannot be and that instead we should just attempt to represent the human condition in all its varieties and complexities.

Whether or not the science of well-being falls under the naturalist or the interpretivist ideal depends entirely on how the options are carved out. It is possible to make naturalism so inclusive that only utter sceptics would end up as interpretivists. But, importantly, such a classification will turn out entirely beside the point for our case. We shall see that the five core commitments of the normal science of well-being of today have

---

8. Little (1991) is a representative textbook. Taylor (1971) is a classic twentieth-century exceptionalist manifesto from a tradition going back to German idealism. More recent discussions and textbooks have largely moved on from this debate (Guala, 2007; Risjord, 2014).

both naturalist and interpretivist features. I stress this mixture as a way of exposing the diversity of the enterprise.

Commitment 1:Well-being is valuable. A central tenet of naturalism from at least as early as John Stuart Mill's (1882) *System of Logic* is value freedom. A social science, just like natural science, should study empirical facts and relations between them. The choice of which facts to study will be value-driven, as Max Weber (1949) allowed, but this is consistent with leaving recommendations to policymakers.[9] When the object of science has an apparently inescapable normative content, a naturalist would normally insist on separating the normative from the descriptive content, keeping only the latter as part of science and relegating the former to ethics and politics. I evaluate these proposals in detail in Chapter 4, arguing that they are a bad idea. For now I just point out that by and large the science of well-being does not follow the naturalist's advice: normative claims, albeit not always explicit and satisfying to philosophers, are part-and-parcel of the science of well-being.

One example is the debate about how to conceptualise and measure the well-being of a nation. It is motivated by a perceived failure of purely economic indicators such as the GDP and gross national product (GNP) to capture the state of communities. Among the inspirations are the Easterlin Paradox as well as Bhutan's pioneering Gross National Happiness Index. There is no shortage of academic opinions on the proper replacement, or complement, of these standard economic measures. Daniel Kahneman and Richard Layard among other psychologists and economists have advocated a hedonic measure—a nation is doing well to the extent that its populace has on average a favourable balance of positive over negative emotions (Kahneman et al., 2004b; Layard, 2005). Development economists typically favour measures based on consumption, access to resources, and other

---

9. 'A scientific observer or reasoner, merely as such, is not an adviser for practice. His part is only to show that certain consequences follow from certain causes, and that to obtain certain ends, certain means are the most effectual. Whether the ends themselves are such as ought to be pursued, and if so, in what cases and to how great a length, it is no part of his business as a cultivator of science to decide, and science alone will never qualify him for the decision' (Mill, 1882, Chapter 12 of Book VI).

objective indicators (Dasgupta, 2001; Deaton, 2016; Nussbaum & Sen, 1993). Yet others opt for life satisfaction, a metric that, it is claimed, best reflects individuals' evaluation of life (Diener et al., 2008). The opponents often do not hide their normative disagreements. George Loewenstein (2009), an eminent economist who raises worries about the purely hedonic measures, titles one of his contributions 'That Which Makes Life Worthwhile'.

True, some research in this field can proceed relatively value-free— take a range of those emotions that people call positive, describe the causal network that surrounds them, and do not say anything specific about their normative status. Positive psychologist Martin Seligman takes that route in a *New York Times* interview:

> My view of positive psychology is that it describes rather than pre-scribes what human beings do. . . . I don't want to mess with peo-ple's values. I'm not saying it's a good or a bad thing to want to win for its own sake. I'm just describing what lots of people do. One's job as a therapist is not to change what people value, but given what they value, to make them better at it. (quoted in Tierney, 2011)

No doubt this is one route and scientists sometimes take it. But note two facts. Which emotions and activities to pick out as potentially rele-vant to well-being is not a value-neutral choice. This is true whether or not scientists demure from spelling out the relationship of these states to well-being. Second, this supposedly modest route is not typical. Often scientists are more ambitious than this. They wish to know whether and which positive emotions are *good for us*: how they enable better func-tioning both at individual and community levels (Fredrickson, 2001), but also whether they harm us sometimes (Gruber et al., 2011). In referring to 'better functioning' and 'harm' these researchers presup-pose a notion of well-being, and this is where the substantive normative assumptions enter.

When making room for values in the definition of the object of study, the science of well-being is rejecting or at least amending a core com-mitment of value freedom, a thesis that Hugh Lacey (2005, pp. 25–26) called *neutrality*. According to it, scientific claims should neither presuppose nor support moral or other value judgements. Though this does not prevent the science of well-being from being value-free in other senses, its rejection of neutrality, even if not universal, is a notable and an antinaturalist feature.

Commitment 2: Well-being claims are generalisable. A major goal of the science of well-being is the development of more or less general causal models of determinants and risk factors of well-being at biological, psychological, organisational, and broader social levels. In embracing this goal, scientists apparently reject the idea that well-being is an idiosyncratic personal phenomenon that does not admit of population-level analysis. Instead the science of well-being operates on the assumption that the social world has causal laws or at least generalisations that could play the role of laws. These laws do not need to apply to all humans at all times and places. They may hold only at the level of community or individuals in specific circumstances (to wit caretakers of the chronically ill, poor single mothers in the United Kingdom, refugees). The generalisations in question usually relate well-being to a socioeconomic or psychological variable such as unemployment or a personality trait, or an activity such as volunteering or commuting. These generalisations are discovered empirically following qualitative or quantitative methods. The science of well-being at this point is a field science, rather than a laboratory- or a model-based one.

In pursuing this commitment the science of well-being rejects two pillars of interpretivism: that the social world is too complex (or too open, or too free, etc.) for any meaningful generalisations and that social explanations should be couched primarily in terms of reasons not causes.

Of course, it is one thing to be committed to this goal and another to actually find such generalisations. Does the science of well-being have any successes to show?

One issue that has occupied researchers and captured public imagination is the stability, or lack thereof, of self-evaluations of well-being. In question is the alleged human ability to adapt, that is, to regain previous levels of subjective well-being, to what seem huge changes in circumstances, such as winning the lottery or losing mobility. To explain this effect some have proposed the *set point theory*—genes and early environment give us a range of happiness to which we invariably return after perturbations. A good example of progress in testing generalisations is the recent updating, even debunking, of these early claims. It turns out that adaptation has a fairly restricted domain and a variable pattern across people. Divorce, serious disability, and unemployment are very hard to get over, while adaptation to the death of a spouse is long but doable (Lucas, 2007).

What about causality? Although notoriously difficult to infer from observational data, standard techniques such as randomised controlled trials and instrumental variables are entering well-being research too. One recent randomised controlled trial examined the effect of job training and supplemented income on a group of poor single mothers in the United Kingdom. The findings are clear and unexpected: their subjective well-being was lowered by greater professional expectations and greater earning power (Dorset & Oswald, 2014).

Commitment 3: The experience of well-being matters. Philosophers may disagree on whether experience directly constitutes well-being (according to hedonists) or merely contributes to it contingently (according to others), but in the sciences the implicit consensus is that studying well-being requires studying experience. The search for causal generalisations coexists with genuine concern with what well-being (or ill-being) *feels like* and how it is understood by the subjects. The classic interpretivist goal is understanding the meaning of actions, the content of experiences, and inscribing those in 'thick descriptions'. It is fair to ascribe to the science of well-being *some* form of such a commitment, though it is realised in very different methodologies.

On the quantitative end, this commitment takes the form of questionnaires or experience sampling. Formal questionnaires or scales, as we shall see later, are the main method for reconstructing and measuring various aspects of well-being using the reactions of subjects to the items comprising these scales. These questionnaires range from gauging a person's feeling ('How anxious do you feel?'), to gauging their judgements ('Is your life going well according to your priorities?'), or their perception of facts deemed important ('Do you feel in control of your circumstances?'). They can be longer or shorter, structured or free, and administered through various media. Some well-known questionnaires include the Satisfaction with Life Scale (Diener et al., 1985), the Positive and Negative Affect Scale (Watson et al., 1988), and the Nottingham Health Profile (Hunt et al., 1981), which measure respectively life satisfaction, happiness, and health-related quality of life.

Experience sampling, on the other hand, aims at detecting and recording the many facets of experience as it is happening. Going through their day, subjects are prompted by a beeper to rate themselves on a variety of positive and negative emotions, their quality, intensity, and so on. Out of these ratings there emerges a picture of how the person felt as time went on and their circumstances and activities changed. Recently, using this method Kahneman and his coauthors (2004a) have studied the daily experience of Texas women who famously found taking care of children to be less pleasant than even housework.

On the qualitative side there are the old and trusted tools of anthropology and sociology. These include ethnographies and open interviews. Recent examples of explicitly ethnographic research on well-being include studies of refugees, families on welfare, intensive care nurses, and many more. With the rise of cross-cultural studies of well-being, these methods become all the more prominent and important, since it is hard to interpret the meaning of responses to questionnaires without talking to people properly.[10]

Notably, even in projects far removed from the qualitative approaches—for example, inference by economists of preferences from choices—the latest methods have abandoned the scepticism about tapping human experience that characterises the classic economic approach rooted only in behaviour. There is growing recognition that only some preferences and only some choices can reveal what really matters to people and that to detect these requires a host of psychological and cultural knowledge, and perhaps even talking to people.[11]

---

10. These themes are explored in Diener and Suh (2000), Camfield et al. (2009). For studies on refugees, families on welfare, and intensive care nurses see respectively Kopinak (1999), Chase-Lansdale et al. (2003), and Einarsdóttir (2012).
11. See Appendix B on the economic sciences.

Commitment 4: Well-being is measurable. 'If you treasure it, measure it', announced Sir Gus O'Donnell in his presentation 'Well-Being Statistics: How Will Whitehall Respond?' delivered on November 2, 2011, in front of the All-Party Parliamentary Group on Well-Being Economics in Westminster. The 'it' was well-being. At the time he was an outgoing cabinet secretary, the highest official in the British Civil Service, and his speech underscored the embrace of the new science of well-being by the UK government. Central to this embrace is measurement. Measures of well-being were to be taken by the Office of National Statistics, of which we will hear in Chapter 4. But even more significantly the far less adventurous UK Treasury dominated mostly by trained economists agreed to mention subjective well-being in its official guide to cost-benefit analysis, the Green Book.

Unsurprisingly, the scientists—some of who revolve in these circles too—are equally confident in their ability to measure well-being. For them the question at this point is not whether well-being is measurable. Their bet is that it is, and the debate has moved on to the plusses and minuses of specific measures. The sceptical view—that well-being is not the sort of thing that can be measured—is still live, naturally among the critics of the science. I examine one such argument, by Dan Hausman, in Chapter 5. But I focus on it as a philosopher, because studying this sceptical position reveals fundamental assumptions of this science. It is, however, not an argument that worries many scientists. For them the measurement project is marching on, largely in accordance with the standard psychometric procedures for developing valid measures. These procedures, my focus in Chapter 6, produce large databases of already validated questionnaires and, for those who insist on creating new ones, step-by-step instructions on how to do so.

When controversies arise, they do so in regard to specific measures, for example, judgements of overall satisfaction with life. There is a longstanding concern with their alleged fickleness: apparently finding a coin, or seeing a person in a wheelchair, or being reminded of the weather, can drastically change a person's evaluation of their well-being. These effects spawned both explanations of how these judgements are formed (perhaps they are constructed on the spot and deeply susceptible to mood) and also attempts to probe their replicability. The latter, however, reveal that judgements of life satisfaction are far more robust than initially claimed, so much so that the weather/coin/wheelchair effects that so excited scholars and the public just a few years ago cannot be replicated (Lucas, 2013). The context in which people are asked to judge their life satisfaction—what they are thinking at the time and in what circumstances—clearly affects this judgement. But whether these context effects make these measures unusable and uninformative is far less clear.[12] So their widespread use continues.

Measurement is a quintessentially naturalist ideal that goes hand-in-hand with Commitment 2 to produce general claims about well-being and with the quantitative wing of Commitment 3 to study subjective experience. Once well-being is treated as a measurable quantity, it can be plugged into generalisations that describe how a given level of well-being as it is experienced depends on a given variable. What about the use of this knowledge?

Commitment 5: Well-being science has applications. It takes all four pillars of normal science to support the fifth. This enterprise wears its policy, medical, business, and activist aspirations on its sleeve. Well-being has become an economic resource and a business tool, a development reflected in the rise of 'corporate wellness programs', life coaches, consultancies, and an intense data-gathering effort about the emotional state of employees and consumers. On the activist side, well-being findings are often recruited to tell us what is wrong with the way middle-class Westerners live and with what they value; from isolation, to consumerism, to the medicalisation of grief and sadness. This is how a domesticated version of Buddhist techniques such as mindfulness-based stress reduction entered both self-help, positive psychology, and mainstream medicine.[13] The science of well-being speaks to governments too, slotting itself naturally into evidence-based policy movement, endeavouring to show which policies, therapies, interventions,

---

12. The original findings are in Schwarz and Strack (1991, 1999). See Deaton and Stone (2016) for the latest evidence of context effects and Lucas et al. (2016) for a defense.
13. For a critique of modern life from this point of view see Haybron (2008, Chapter 12), among other places. For a classic of positive psychology see Seligman (2004). For a history of mindfulness in North America see Wilson (2014).

and community arrangements most efficiently relieve suffering and improve the well-being of all concerned. The triumphs of the activist scientists include the establishment of well-being indices; systematic data gathering and reports; incorporation of mental health initiatives into schools, hospitals, armies, welfare systems, and many more other such plans.[14]

I have listed five commitments of the science of well-being. The last of them, policy hopes, has historic associations both with naturalists and, via critical theory, interpretivists, so we have a draw: three points (generalisability, measurement, and policy aspirations) for naturalism and three points for interpretivism (value-ladenness, focus on lived experience, and policy hopes). In this sense the science of well-being is mixed. It has goals and methods typical of both interpretivist and naturalist ideals.

We could note further features of this mixedness. Mathematical modelling and the elaboration of abstract theory, so important to economics, physics, and parts of biology, have not arrived to well-being. Empirical studies of large- and small-scale causal networks that are widespread in epidemiology, econometrics, and climate science are, by contrast, underway. The science of well-being inherited controlled experiments and psychophysical measurement from psychology, but these do not define it. Instead it is more explicit in its value-ladenness, more friendly to anthropological methods, and more humanist at least in its official aspirations.

This does not make the science of well-being unique. Health and climate science have mixed features too. Indeed the categories of social versus natural science, interpretivism versus naturalism, ideographic versus nomothetic methods, may or may not retain relevance for new hybrid disciplines such as this one. The philosophy of the science of well-being is not a branch of philosophy of social science, nor of philosophy of natural science for that matter. So what will it be?

---

14. For the rise of official well-being statistics see Stiglitz et al. (2010), Office of National Statistics (2012, 2013), Self et al. (2012), OECD (2013), Kahneman et al. (2004b). For their policy relevance see Diener et al. (2008), Huppert et al. (2003), Dolan and White (2007) among many others.

INTRODUCTION

AN AGENDA FOR PHILOSOPHY

This book's title promises *a*, rather than *the*, philosophy, so I start by mentioning some agendas I am not pursuing.

Mine is not an exercise in political theory. As Commitment 5 illustrates, science of well-being is often driven by a kind of welfarism—a view that well-being should be a goal of public policy. I do not defend or criticise welfarism here, because strictly speaking the pursuit of knowledge about well-being does not depend on the truth of welfarism—but only 'strictly speaking'. In reality it is hard to imagine anyone bothering with this science if well-being was not a relevant policy consideration. So I assume that much and turn to the proper shape of such knowledge, without weighing in systematically on how this knowledge should be used by polities, democratic or otherwise.

I also disavow the goals of either debunking or vindicating this field wholesale. Critiques of sciences such as ours tend to expose them as tools of capitalism, of neoliberal state, of managerial control, a fad, and so on. The enthusiasts, on the other hand, see vision, humanity, and empowerment.[15] There is truth in each perspective, but the scope of the field as I delineate in the five aforementioned commitments is too wide and too inclusive to make either one or the other a plausible full story. There is no one way to generalise, to measure, to respect subjective experience, nor one way to practice well-being activism. Because of this diversity, neither debunking nor defense are appropriate. In places I help myself to ideas of each camp, but the moral case that properly considers the promises of this science against its dangers will be complex, and I do not endeavour to present it fully.

Finally, mine is *a*, rather than *the*, philosophy in another sense. A comprehensive philosophy of this science would cover a great deal of territory just because it raises many of the very same questions as other field sciences: how to infer causes and to measure their magnitude, how to strike a balance between generality and specificity of theories, how to use first-person reports, how to elicit phenomena without distorting

15. For pessimism see Davies (2015), Rose (1990, 1998), Lazarus (2003). For optimism see note 2 and responses to Lazarus in a special issue of *Psychological Inquiry*, 14(2), 2003.

them, how to confirm hypotheses without misusing statistics, and so on. I do not discuss these worthy issues.

My gaze is selective but also worthy. I set myself one question that no philosophy can ignore: *How can the science of well-being produce knowledge that is properly about well-being?* Since such knowledge would be laden with apt values, I refer to it as the Question of Value-Aptness.[16]

When a headline proclaims that a happy marriage requires a wife slimmer than the husband[17], I need to know what these researchers mean by 'happy marriage' and whether it is indeed good for me, before I rein in my appetite. Less frivolously, much of the methodology of the science of well-being rides on how we answer the Question of Value Aptness. Three issues do in particular:

1. How well-being should be defined in a given scientific project.
2. How well-being should be measured.
3. How the science of well-being can retain objectivity in the face of values.

This is my, admittedly selective, agenda for a philosophy of the science of well-being. Each chapter in this book addresses some part of this agenda. But before I say more we need to see why the Question of Value Aptness is far more taxing than it seems.

An intuitive approach to value-aptness is this: the science of well-being is value-apt to the extent that the value-laden concepts that feature in its claims are appropriately informed by the best existing normative theories—in this case normative theories of well-being. Unfortunately, this answer is not so much wrong as very uninformative. The hard part is to specify, first, what 'appropriately informed' means and, second, by which normative theories.

The philosophy of well-being as practiced today is a study of what makes a life or some part of it good for one. Philosophers in the

---

16. The expression 'value aptness' was first voiced to me by Stephen John.
17. http://www.telegraph.co.uk/news/8646930/Happiness-is-based-on-wife-being-slimmer-than-husband-according-to-study.html

analytic tradition call this value 'prudential' and distinguish it from moral, aesthetic, epistemic, and other values. It is a pursuit with a much longer history, albeit a less public present, than the science of well-being. Because philosophers describe their goal as the articulation of theories of well-being, it is a natural place to turn for our value-aptness fix. Philosophers are interested in defining well-being, scientists in measuring it, so a division of labour suggests itself: let the philosopher tell the scientist the values that the measures are supposed to capture.

Alas this proposal for a division of labour is doomed from the start. The science of well-being should not seek out philosopher-kings—the definitions of well-being usable in the sciences must be sensitive not only to the normative theories of the good life but also to the practical constraints of measurement and use of this knowledge. But the goals of theorising about well-being in philosophy as it is currently practiced are not sensitive in this way.

OBSTACLES TO VALUE-APTNESS

Before we say any more we need a crucial three-way distinction between *theories, constructs,* and *measures* of well-being. Very roughly, theories are the preoccupation of philosophers, constructs and measures of scientists. A theory of well-being is a study of well-being's essential properties, those that make it well-being rather than something else. Philosophers often do this by attempting to specify necessary and sufficient conditions for classifying a person as 'doing well'. The term 'construct', on the other hand, is used mostly by psychologists and is just another name for an attribute or a phenomenon, in our case the state of well-being in the subjects of a scientific study. Constructs are usually unobservable but have various observable manifestations. For example, those who do well are less likely to commit suicide. Finally, measures are ways of eliciting the observable indicators of constructs. For example, a score on a questionnaire might be such an indicator. If this questionnaire is really good at detecting well-being, it is said to be a valid measure of this construct.

Ideally, theories, constructs, and measures should stand in the right relation to each other. Measures must reliably track constructs, and our choice of constructs must be properly informed by theories. I alluded that the theories of well-being from philosophy are not capable of properly guiding the development of constructs and measures in science. Why not?

We might be tempted to blame it on the simple fact that philosophers disagree about the nature of well-being. Over the last two millennia, they have proposed and developed several theories of well-being, most notably a number of variations on the original ancient proposals of eudaimonism and hedonism, plus several on the more recent desire-fulfillment view. Appendix A offers an overview of these theories. For now, we use a basic distinction between *subjectivists* and *objectivists*. Subjectivists insist that nothing can be good for us unless we desire, prefer, or endorse this good. The objectivists disagree: a loving relationship or positive emotions, for example, are good for us whether or not we want it. The main version of subjectivism takes well-being to consist in the fulfillment of a person's deepest and most important desires, goals, or values. Most objectivists about well-being insist on the fulfillment of human nature or flourishing, adopting a version of eudaimonism going back to Aristotle and other Greeks. Some objectivists are hedonists for whom well-being consists in a life of positive experiences. Philosophers have naturally found counterexamples to each theory, that is, made-up scenarios that fit the theory but intuitively do not count as well-being (or the other way around). At this point, the philosophical literature on well-being is extensive, and each of the major options have grown elaborate and intricate under the weight of counterexamples. However, there is no consensus: not on whether well-being is wholly subjective or not, not on what exact mental states are partially or wholly constitutive of it, and not on the level of those states that is necessary. Instead a variety of different answers to these questions coexist in the literature.

But deep philosophical debates in themselves should not stop the study of well-being in its tracks any more than the chasm between empiricists and rationalists about the nature of knowledge stops any other inquiry. Besides, the debates in philosophy of well-being are not normally about *which* goods are prudentially valuable but rather about the *reasons* why they are valuable. So philosophers might all easily agree that pleasant experience matters, success in personal projects matters, living within one's limits matters, and possibly more. This level of agreement could potentially be enough for answering the Question of Value-Aptness.

Rather, the real obstacle to value-aptness is that current philosophical theories are just not *about* the right thing. They are about a concept of well-being *in general, all things considered*, the sort of concept we use when we evaluate either a life as a whole or a period of life in all its prudential aspects. This is a very important context but also a fairly narrow one.

Take the question: 'How is Mo doing?' This question might be asked in two kinds of contexts: a general and a specific one. A general context considers Mo's life as a whole, or his current state at a time all-things-considered. Say Mo's close friend asks him 'How are you?' in that significant tone of voice in a heart-to-heart conversation, or 'How did Mo's life go?' at Mo's funeral. This is a context in which we must take account of all the important things in his life (either up to then or as a whole), evaluate how he is doing or has done on each account, and then aggregate all the important elements to produce an overall judgement. This is what I mean by general evaluation. If, on the other hand, Mo hears 'How are you?' from his family doctor at an annual checkup, the same question invoked a context-specific evaluation—'Are you feeling healthy?' This would be a contextual evaluation—only a particular aspect of well-being is in question here. Contextual evaluations also aggregate some information but not as much as general ones. Still there is a difference in degree.

Some scientists of well-being are interested in the all-things-considered well-being—positive psychologists write books on how to improve one's life. But more often than not the sciences dwell in the contextual territory. Researchers ask how a person or a group of people are doing *given* their circumstances and *given* the special focus these researchers adopt. A therapist is interested in how her patient is recovering from depression; a social worker in whether his clients are managing to rebuild their lives after a crisis; a team of development economists are interested in a community's access to basic goods.

Philosophers have typically theorised only about the first kind of well-being—the agent's overall all-things-considered well-being, not the second kind. A hedonist philosopher will take well-being to consist in *all* the pleasures. A desire theorist, once they have identified the set of desires that are well-being relevant (see Appendix A on various restrictions), will identify well-being with the fulfillment of *all* these desires in their order of overall importance, and so on. This generalist focus persists whether philosophers talk about *temporal well-being* (well-being at a specific point in time or a period) or *life well-being*. Even those philosophers willing to entertain the idea that the notions of life and temporal well-being obey different rules[18] theorise about the most general evaluation. Context is still absent, or rather it is present but only the one general context.

This is, of course, perfectly fine. There are virtues to focusing on all-things-considered well-being. It is the human condition! But this focus is not adapted to the Question of Value-Aptness because this question calls for translation from the general to contextual evaluation. The unique focus on general well-being puts the philosophical project at odds with the project of the sciences. It leaves current philosophical theories of well-being far less relevant for science. There are no ready-made theories for scientists to take off the philosopher's shelf. For all their internal intricacy and sophistication, these theories are not intricate and sophisticated enough to serve where help is most needed, that is, in the selection of constructs and measures. And the tragedy is not just philosophy's: for if there is no proper value-based justification for construct development, it follows there is no justification for the knowledge claims of the science of well-being. Everybody is worse off—philosophers, scientists, and the users of science.

This is why the Question of Value Aptness will not be settled merely by bringing existing philosophy into the picture. Rather we need to start practicing science and philosophy in a joined up manner.

---

18. 'Temporal well-being' is the expression used by Broome (2004, Chapter 6), 'life well-being' is Kagan's (1992). Velleman (1991) and Kauppinen (2015) spell out the specialness of life well-being.

CONSTRUCT PLURALISM

Indeed, when we look at the sciences of well-being we see a great variety of contextual definitions and measures. Psychology alone boasts three approaches to defining and measuring well-being, economics two, and projects in the policy and clinical sciences yet more. Some definitions represent only the subjective judgements of people about their own lives, others contain objective quality of life elements; some are based only on the subjects' affect or emotions, others on their cognitive judgements, and so on. As a result, many different things are called 'well-being'. Constructs said to represent well-being in gerontology and medicine differ strikingly from those in development economics and child psychology; they can even differ substantially within different subfields of the same research area. What I call *construct pluralism* is a pervasive and manifest feature of the science of well-being, a fact I summarise in Table I.1.

Each row represents an area of science that uses a notion of well-being. The columns aim to give, respectively, a philosophical theory commonly assumed by this area of research (Column 1), the constructs built on the basis of this theory (Column 2), and the measures that are supposed to capture the construct (Column 3). Appendix B gives the necessary background and references. Notice that in some rows I left the theory column blank. Why? Because in these areas researchers use a context-specific, not a general, notion of well-being, and it is often not clear what philosophical theory is supposed to justify the choice of construct. But the problem is bigger than it looks. Why are there three different theories in the first four rows? Is each one of them equally necessary? Isn't there one correct theory of well-being?

Construct pluralism presents us with two tasks. The first is methodological: which of the many things called 'well-being' in the sciences is the correct construct to use and for which purpose? Any philosophy of the science of well-being worth its salt must come with recommendations for how researchers should choose their constructs. The second task is philosophical—to explain why science lives with pluralism while philosophers search for a single correct theory of well-being.

Table I.1 CONSTRUCT PLURALISM

	(1) Theory	(2) Construct	(3) Measure
Psychological sciences	Hedonism	Average affect	Experience sampling, U-Index, Positive and Negative Affect Scale, SPANE, Subjective Happiness Scale, Affect Intensity measures
	Subjectivism	Subjective satisfaction	Satisfaction With Life Scale, Cantril Ladder, Domain Satisfaction
	Eudaimonism	Flourishing	PERMA, Psychological Well-Being Index, Flourishing Scale, Warwick and Edinburgh Mental Wellbeing Scale
Economics	Subjectivism	Preference satisfaction	GDP, GNP, household income and consumption,
Development sciences	Objective list theory	Quality of life	Human Development Index, Dasgupta's index
Policy sciences	Pragmatic Subjectivism (Haybron and Tiberius, 2015)	National well-being	UK's Office of National Statistics Measure of National Well-being, Legatum Prosperity Index, Social Progress Index, OECD Better Life Index
Medical sciences		Quality of life under various medical conditions	Nottingham Health Profile, Sickness Impact Profile, World Health Organization Quality of Life, Health-Related Quality of Life, QUALEFFO

Table I.1 CONTINUED

	(1) Theory	(2) Construct	(3) Measure
Child sciences		Child well-being	US Department of Health and Human Services Children's Bureau Child Well-being Measure (three domains of assessment—family, education, mental health and physical needs); UNICEF's State of the World's Children; parental evaluation; Stirling Children's Wellbeing Scale, and other scales.

Of course the two tasks are related. Depending on whether and how construct pluralism is justified philosophically, the method for fitting constructs to projects will be different. Part I of this book tackles philosophy, while Part II examines the implications of this for the science. However, the philosophy is not in the driver's seat here. I take construct pluralism to pose a genuine objection to the philosophical status quo that proceeds on the assumption of there being a single correct theory of well-being. In Chapters 1 and 2 I accordingly propose a revision of the philosophy of well-being. But philosophy is not purely a passenger either. No choice of a given construct of well-being is intelligent and justified without a theory underpinning it, and building such theories is a distinctly philosophical exercise. It just will not be the sort of theory that philosophers are used to.

A REVISION OF PHILOSOPHY

In Part I I develop a philosophical view called Well-Being Variantism, according to which there is neither an all-purpose concept, nor an all-purpose theory, of well-being. Instead there are (a) several different concepts that are appropriately referred to as 'well-being' and (b) possibly also several substantive theories that describe the referent of these concepts in different contexts.

In Chapter 1 I explore the first thesis, borrowing ideas from epistemology about how claims and attributions of knowledge depend on context. I favour a version of contextualism according to which the semantic content of well-being expressions changes with the context in which it is asserted. In some contexts well-being means all-things-considered evaluation and in others a more limited judgement about certain specific conditions of life. This variability is not the full explanation for construct pluralism—their variety is also due to substantive disagreements about what well-being is and to pragmatic choices about what each research project is best positioned to measure. But instability of meaning is part of the story.

Contextualism is only about the content of well-being claims and in this sense a fairly tame thesis. By itself, it does not yet imply that there is no single usable substantive theory that regulates the referents of each contextual well-being notion. But construct pluralism could lead us to consider this stronger possibility too. In Chapter 2, I articulate this second variantist thesis—that the master theory of well-being is not forthcoming nor indeed needed.

Such flirtation with pluralism will come as no surprise to many philosophers of science. They have learned to temper their expectations about the power of theories as opposed to more localised sources of knowledge such as models, mechanisms, and instruments. I think philosophers of well-being should do likewise. On the standard view once common in philosophy of science, for any particular phenomenon in need of representation, a corresponding theory should be able to imply this phenomenon given certain assumptions. This is the *vending machine* view of theory, to borrow Nancy Cartwright's (1995, 1999) apt term. Chapter 2 argues that philosophical theories of well-being are not vending machines. We just do not have such powerful theories of well-being, and if we held the empirical study hostage to the vending machine view then such a study would never get off the ground. Instead the role of philosophical theory is different: it is to assemble a *toolbox*, again Cartwright et al.'s (1995) term, full of concepts that help in developing any number of constructs and measures.

ENTER MID-LEVEL THEORIES

How are scientists to choose the right construct of well-being for their project if not by relying on a master theory of well-being?

Call the standard philosophical theories of well-being—hedonism, subjectivism, and eudaimonism—the Big Three. The Big Three are *high* theories in that they are about persons in the broadest possible sense without any specific context. I propose to distinguish high theories from *mid-level* theories. Mid-level theories are about the well-being of *kinds* of people, often groups, in *kinds* of circumstances: children, children in the welfare system, former child-soldiers, working mothers, caretakers of the ill, post-Brexit Britain, and so on. These kinds can be as general or as specific as our scientific and policy projects require. Mid-level theories are about the conditions of actual flourishing of these kinds given their environments.

They are mid-level because they are in between the high Big Three and the very specific measures of well-being in practical and scientific contexts. To be sure, a mid-level theory depends on high theory, but the two do not fully share criteria of assessment. If the goal of a high theory is to systematise as many disparate judgements about well-being as possible into a maximally simple consistent and yet powerful set of propositions, a mid-level theory need not necessarily. It systematises some but also has goals of its own, most importantly to enable and guide social measurement and application.

Where do mid-level theories come from? Implicitly they already exist. It is an implicit mid-level theory that motivates specialists on child well-being to attend to play and attachment, while specialists on national well-being focus on the sustainable use of resources, to use two examples. But these theories are often not well worked out and not well connected to measurement, policy goals, or the Big Three. They need to be for construct pluralism to be justified.

I conceive of the relationship between high theories, mid-level theories, constructs, and measures to be as depicted in Figure I.1.

High theories *inspire* mid-level ones by providing conceptual tools that enable the latter's formulation. Mid-level theories *justify* different constructs of well-being, whereas different scales enable *measurements* of the constructs. The arrows are different in each case because the relations are different. To inspire is not to justify and it is not to measure. Note also that a single high theory can inspire two different mid-level theories (or none at all). A single mid-level theory can justify several constructs, and a single construct can be measured by several scales or not have a measure at all.

Mid-level, not high, theories occupy center-stage in my proposal. They enable the science of well-being to be value-apt, and they are a

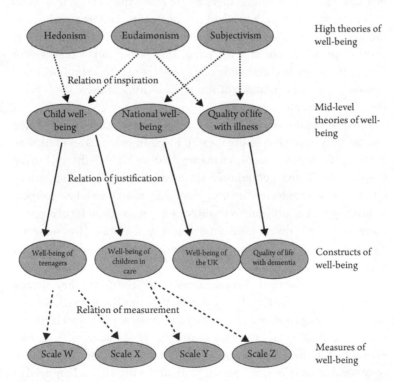

Figure I.1

far more urgent task than another high theory no matter how intricate. This is not to say that the classic philosophical chestnuts about well-being are irrelevant to science. Is well-being just a mental state? Are some mental states more valuable than others? Can a knave fare well? These questions do regularly come up, and sometimes the application of scientific knowledge about well-being requires taking a stand in these controversies. But they are not relevant as often as the extent of the philosophers' attention to them seems to indicate. When they do arise they can rarely be resolved by appeal to high theories but more often instead by appeal to an implicit mid-level theory, pragmatic considerations, or else by political means.

PUTTING PHILOSOPHY TO WORK

In Chapter 3 I show how a mid-level theory can be built. My collaborator on child well-being, public health scientist Ramesh Raghavan, taught me that child well-being is an area of intensive scientific study and policy interest. The notion of 'best interests of the child' is central to welfare policy in many nations. Measures of child well-being range from the most basic (used by welfare agencies) to the more refined (used by UNICEF, charities, and child development specialists). Yet the theoretical question 'What is child well-being?' has so far received no rigorous answer from either scientists or philosophers. The existing theories of well-being, with the exception of hedonism, are about the ideal rational adult. And hedonism, which is about animals in general, poorly captures the importance of growth, exploration, and development so unique to children.

Rather than being derived from general theories of well-being, child well-being needs a distinct substantive theory of its own, which can be used to build locally appropriate constructs. Such a theory needs to be based on empirical knowledge about children and their development, as well as on a philosophical conception of what it is to be a child. A high theory can serve as a constraint, but it does not imply a theory of child well-being. Using these various sources of knowledge I put together an account that sees child well-being as responsive to two demands: a forward-looking one that sees childhood as a step toward adulthood and a present-looking one on which childhood has value in and of itself.

Other neglected and much needed mid-level theories could be well-being for people with specific disabilities, traumas, or chronic illnesses; well-being of the displaced and the refugees; well-being of caretakers, and so on for however many social kinds need a theory of well-being. Ideally a comprehensive philosophy for the science of well-being would include a map from contexts of research to corresponding mid-level theories and then to corresponding constructs. In this book I do not offer such a map simply because I do not know enough about the specific challenges and achievements that each of the different contexts bring. Building a mid-level theory of well-being is hard work as it involves working both from below—the existing empirical base—and from above—the relevant high theories, and then synthesising the two as Chapter 3 illustrates. But I hope to lead by example. Mid-level theories are badly needed, and philosophers who are not averse to learning facts on the ground have the perfect set of skills to build them.

While Part I is concerned with answering one part of the Value Aptness question, that is, the choice and justification of constructs of well-being, my variantism and mid-level theories do not in themselves provide an answer to whether these constructs can be legitimate objects of science. This is why in Part II I take up classic issues in philosophy of science—objectivity and measurement. Here I am concerned with showing when and how constructs of well-being can be not just well-grounded theoretically but also obey constraints of scientific method.

Chapter 4 asks: What is objectivity? Commitment 1 of our normal science notes various grades of value-ladenness in this enterprise. One of these grades—the use of normative assumptions in the definition of well-being—raises a worry about objectivity of this science. Value freedom has been an important ideal dedicated to guarding science from bias and wishful thinking. I offer a conception of objectivity appropriate for this case and indeed other sciences that deal with normative concepts. Objectivity does not imply handing over decisions about values to policymakers and other users of the science. We should not try to eliminate the normativity so essential to the constructs and measures that the science of well-being uses. If value freedom requires this elimination, so much the worse for value freedom. Normativity in itself does not make the science of well-being dangerously political and ideological—not unless this normativity is used to impose objectionable values on the unsuspecting users of this science. But it does not have to. To be objective the science of well-being has to be based on values that are out in the open and to the extent possible vetted by a deliberative process. This ideal is neither impossible nor problematic. The science of well-being can and should strive to be objective in this sense.

But there is another sense of objectivity that this field aspires to, as does any other science—that is to measure what is really there. Whether well-being is measurable and how much we can trust the current measures is the focus on the last two chapters. In Chapter 5 I discuss what I take to be the most compelling argument for scepticism about well-being science—that it aspires to measure something that is too diffuse, too personal, and inherently unmeasurable. Put forward recently by Dan Hausman (2015), this argument rejects the 'normal' modus operandi I sketched out earlier, proposing that none on the existing measures respects what to him is a nonnegotiable feature of well-being. This feature is that well-being, no matter which of the Big Three is endorsed, is a value that aggregates goods in a way that respects individual identity. This aggregation must be holistic and sensitive to values and circumstances, and it is precisely this heterogeneity of well-being that any measure that purports to apply to masses of people is bound to miss.

I have a great deal of sympathy for Hausman's argument. Indeed, I think it establishes convincingly that the science of well-being is unlikely to be a science of *individual* well-being in the all-things-considered sense that earlier I identified as the sense to which philosophers have been exclusively attending. Chapters 1 and 2, however, reject this sense as unique and uniquely interesting. When it comes to contextual well-being, which science typically predicates of kinds rather than of individuals, Hausman's scepticism is less warranted. The uniqueness of any particular earthquake does not undermine the possibility of a science of seismology. Nor should the uniqueness of a particular good life undermine the science of well-being.

In mounting this response I also appeal to an aspect of Hausman's case that is representative of other critics of the science of well-being—that is, rejecting existing measures, be they questionnaires or indicators, on intuitive grounds. How could, such critique goes, questionnaires ever manage to capture this or that aspect of well-being? I take a dim view of such arguments. To show that a measure is invalid it is not enough to list plausible ways in which it might fail or even does fail on occasion. This is because measures of well-being go through a process of validation—most commonly psychometric validation. Since this process is supposed to ensure the validity of these measures, to criticise any measure effectively one must criticise this process.

Psychometric validation has been almost entirely ignored by philosophers. But no serious discussion of the science of well-being can afford to do so, which is why I devote the rest of Chapter 5 and the whole of Chapter 6 to reconstructing and evaluating the logic behind this enterprise. In my view psychometric validation is based largely on a sound principle that a measure should only be declared valid if its behaviour coheres with the background theory of the phenomenon that this measure tracks. It is the application of this principle that I find lacking. What counts as relevant background knowledge in psychometric validation is too narrow. True to its operationalist heritage, the procedure excludes knowledge about values and relevant philosophical considerations about the nature of happiness, well-being, quality of life, and related concepts. Too often and too mechanically psychometric procedures commit the sin of *theory avoidance*. Scientists are eager to validate their measures against empirical data but not against philosophical theories, even when they are available. This theory-phobic attitude permeates the practice of the psychometric validation of questionnaires. Psychometrics thrives on the statistical analysis of existing questionnaire data and on checking the correlations with other known facts about well-being. While this approach is partially defensible, it outsources too much theory to statistics. A positive way to describe this status quo is as an understandable reaction to the paucity of usable mid-level theories—a status quo that I am keen to change. A less positive stance is to liken the worship of psychometric validation to techno-cratic expertise taking over an issue that is in fact deeply political and moral. Whether a measure of well-being is valid should not be mainly a technical question.

I am optimistic, though, that the objections I raise are not fatal. There is nothing inherently wrong or impossible in the project that is the science of well-being, nor are there insurmountable obstacles to its improvement. This science does, however, call for a rethinking of what it takes to theorise about well-being and to measure it objectively. My intention is to offer such a rethinking.

A GUIDE FOR READERS

Different readers will engage with different parts of this book. Chapter 1 is as close as I get to discussing traditional themes that occupy philosophers in the so-called core areas of analytic tradition. Readers uninterested in the mechanics of the arguments against a single concept of well-being can safely skip its mid-portions. Chapter 2 is a critique of a prevailing method-ology in philosophy of well-being from the point of view of science and of philosophy of science. Chapter 3 on child well-being is again applied phil-osophy. My goal there is to speak to political theorists and social scientists interested in children. Chapters 4, 5, and 6 focus respectively on objectiv-ity, measurability, and psychometrics. The natural audience are scientists of well-being, philosophers of science, and the users of both. Throughout the book I make references to common trends and arguments in both philosophy of well-being and the relevant sciences. Readers who lack this background can refer to two appendices that summarise what I see as state of the art in those areas.

PART I

TOOLS FOR PHILOSOPHY

He revealed to me a whole new world of joys in the present, without changing anything in my life, without adding anything except himself to each impression in my mind. All that had surrounded me from childhood without saying anything to me, suddenly came to life. The mere sight of him made everything begin to speak and press for admittance to my heart, filling it with happiness.

That day ended the romance of our marriage; the old feeling became a precious irrecoverable remembrance; but a new feeling of love for my children and the father of my children laid the foundation of a new life and a quite different happiness; and that life and happiness have lasted to the present time.

Lev Tolstoy, *Family Happiness*

Tolstoy wrote *Family Happiness* from the point of view of Masha, a young aristocrat and a study for his later *War and Peace* character Natasha Rostova. In the first quote she is 17. It is early summer in the countryside and she is falling in love with her future husband. Her happiness is intense, virtuous, and dreamlike. The second quote is the last sentence of the novel: four years later Masha is wounded by the cruel, yet tempting, high society of Peterburg and Baden Baden; the pure love that connected her and her husband early on is now lost forever, but she has gained wisdom, acceptance, and a new kind of happiness. Tolstoy is exploring a simple idea that each stage of life can have its own happiness, its own way of being good for us.[1]

No theorist of well-being in philosophy would deny this, but neither does any of them today make it their goal to say how to live well at different times and during different circumstances in life. Philosophers are content to define what well-being is as one concept and to specify the states that make it up at the most general level. How these states translate into a good life at particular life stages or under certain circumstances is thought to be an empirical or a personal matter, not a philosophical one. I think that, on the contrary, Tolstoy's concern (though not necessarily his ideas about family happiness) should also be a concern of philosophy, especially if philosophy is to be relevant to science. That is, philosophy should do more than just specify well-being at the most general level. Each chapter in Part I contributes to this claim. I start with a foundational question: When talking about well-being, are we talking about just one thing?

1. Tolstoy is using 'happiness' in the normative sense, not just a psychological one. This is why it is appropriate to use 'happiness' and 'well-being' interchangeably in this case.

Chapter 1

Is There a Single Concept of Well-Being?

Books on well-being normally start by clarifying the concept of well-being. This one, on the other hand, starts by raising doubts about the existence of a stable and unified such concept. Instead its meaning is to some extent changeable and fragmented.

In the Introduction I mentioned the diversity of constructs of well-being in the sciences. Depending on the situation, scientists use different definitions of well-being and they use the term 'well-being' to denote quite different things in different research projects. Table I.1 summarised this diversity in column 2: 'well-being' refers to states of mind in psychology (happiness, satisfaction, or sense of flourishing), to satisfaction of preferences in economics, to objective quality of life in development contexts, to aspects of perceived or actual health in medical research, and so on.

Such diversity and liberality is in part due to the recent explosion in popularity of well-being studies. Today more scientists than before claim to be studying well-being. Furthermore, some instances of construct pluralism indicate substantive disagreements about the nature of well-being—for example, is life satisfaction more important to national well-being than happiness? Finally, construct pluralism can be due to different disciplines concentrating on aspects of well-being they know and understand best. However, construct pluralism is not only an artefact of this recent history of science, disagreements about values, and pragmatic choices of researchers. Even outside of science the question 'How are you doing?' invites thoroughly different evaluations, holding fixed the actual state of the person to whom the question is posed. When the question is posed to a pedestrian who slipped and fell on ice, it is a question about whether or not she is seriously hurt by the fall. When posed by a close friend in that 'significant' tone of voice, it is a question about how one's life is going as a whole. This chapter is a proposal about how to interpret this diversity.

What terms and expressions do I have in mind? In English, as in several other languages, the term 'well-being' itself is rarely used in everyday life. 'How is your well-being?' and 'My well-being has been good lately' are thankfully rare. Instead, well-being is invoked by questions such as 'How are you doing?' or 'How have you been?' or 'How is it going?' and by replies 'I am doing really well', or 'I am fine lately', or 'Average'. I shall assume that often enough these questions and answers are about well-being, all the while recognising that sometimes they are not. In North America 'How are you doing?' is used as a greeting to which no answer is expected. Joey, the character from the 1990s sitcom *Friends*, famously used it as a chat-up line. 'How YOU doin'?' he inquired with a twinkle in his eye. These are clear negative cases. A clear positive case is when the question 'How are you?' is asked in that special tone by a friend of a friend or by a therapist of a patient. Then we are definitely talking about well-being, and indeed something close to the philosopher's all-things-considered evaluation. Between these negative and the positive cases there is a huge grey area.

So far, philosophy of well-being has not been sensitive to this grey area. As I have mentioned already, theorists have traditionally concerned themselves with what is good for a person in general, all things considered. Even when talking about so-called temporal well-being, that is, not about life as a whole, philosophers are still interested in the most general essential conditions for such a good. But this understanding does not match the understanding of 'well-being' and its cognates that we find in the sciences and indeed everyday life: here, unlike in philosophy, 'well-being' is often used for a context-specific rather than a general evaluation of a person's state.

There is currently no account of this discrepancy. Is it an indication of a genuine disagreement between philosophers and others? If not, how exactly are we to understand the behaviour of 'well-being' and the related locutions? I consider three options and argue in favour of one of them:

Option 1: The first possibility is just to deny the significance of the diversity in question. One could claim that this diversity, to the extent that it exists, is a mistake or an instance of linguistic carelessness on the part of those who use 'well-being' outside its proper context. Well-being proper is that general, all-things-considered evaluation that philosophers have been concerned about. Whoever uses 'well-being' and its cognates to refer to economic quality of life, or healthiness, or whatever else scientists call 'well-being' that does not plausibly denote general prudential good, is at best confused (perhaps wrongly attempting to economise on words). Well-being invokes a general evaluation, while quality of life, good mood, healthiness, and so on do not. I call this view Circumscription, because it circumscribes the notion of well-being within a narrow domain.

Option 2: Another possibility is to accommodate the diversity that the Circumscriptionist rejects. Perhaps there is a stable content for well-being expressions (an assumption shared with the Circumscriptionist), but different states realise well-being in different circumstances. I call this the Differential Realisation view. For instance, well-being might consist purely in one's emotional balance when ascribed to a depression sufferer, or in one's access to basic medicine and education in an environment of deep poverty, or in realisation of one's dreams and ideals when we evaluate someone's life as a whole. On this view, the semantic content of the term 'well-being' does not change with each change in the environment. Only the truth-makers of the state 'is doing well' (or 'is not doing well') change with context.

Option 3: Finally, one could adopt Contextualism about well-being. Familiar from recent epistemology and philosophy of language, Contextualism about well-being would maintain that the semantic content of sentences in which 'well-being' and its cognates occur depends, at least in part, on the context in which it is uttered. A developmental economist might just mean something different by 'well-being' than does a clinical psychologist. On this view it is impossible to speak of well-being *simpliciter*. Rather, the content of well-being assertions needs to be indexed to specific circumstances (doctor's visit, poverty relief on country-wide scale, heart-to-heart conversation with a friend, etc.). Since these circumstances will inevitably differ from situation to situation, so will the semantic content of 'well-being'. In one situation it will connote a concern of a doctor for their patient, in another of a social worker for his clients, or of a therapist for her depressed patient, and so on and so forth. The context of an all-things-considered evaluation privileged by philosophers is just that: *one* of the *many* contexts in which well-being is in question.

These three options may not exhaust the space of possibilities, and indeed I briefly mention other options later. There may be ways of combining them depending on the precise features of each. But the options here considered are inspired in part by parallel debates in philosophy of language and epistemology, and, in any case, no other such list is on offer, so the three-way division is a fine starting point.

If these were our options, how should we choose among them? I propose, first, to abandon Circumscription because it is too restrictive about the scope of well-being. We should instead pick between Differential Realisation and Contextualism. Although there are plusses and minuses to both views, there are strong reasons to prefer Contextualism.

1.1. MASHA AND HER 'WELL-BEINGS'

To supplement my examples of construct pluralism in science, here is an intuitive, though not a toy, I like to think, example.

Consider a present-day Masha. She is a heavily pregnant city-dweller walking along an icy street. She slips and lands on her right knee. A Good Samaritan rushes toward her from across the street. 'How are you doing?' he asks her, extending his hand. Masha hesitates: 'I can't walk, but I don't think that anything else is wrong. . . . Can you help me to that bench?' The Good Samaritan does so and leaves after Masha phones her partner who promises to pick her up by car shortly.

That same evening Masha attends a dinner party hosted by a good friend of hers. When they have a quiet moment together, the friend asks: 'So. . . how are you?' Masha is glad at the opportunity to pour her heart out. For starters, her partner's job is temporary adjuncting at a local college and he may well be out of work by the time their child is born. If so, they would have to move again while the baby is still very young, and even that is on the good scenario that he can get another job for next year. Masha gave up her PhD for the sake of following him. Perhaps had she stuck with it, she might have been more successful than him at securing a permanent job. Now with the baby nearly here, her status as a career-less stay-at-home mom is about to be cemented. While she is still happy in the relationship, anxieties, insecurities, and resentments are bubbling up. She has not been sleeping well because of this.

The next day, Masha is visited by a city social worker in charge of supporting new parents-to-be. The social worker quizzes Masha on her income and sources of social support. It turns out that, apart from the partner's salary, Masha can tap into a small fund her extended family set up for her and her future child. Her parents, who live nearby, are thrilled at becoming grandparents, and Masha has always gotten along very well with them. She also has local friends and attends a neighbourhood group for first-time parents.

In this scenario, we witness three judgements that are ostensibly about Masha's well-being. Nothing changes in her life, and yet in each case a different standard of well-being is used. The Good Samaritan has in mind neither flourishing nor positive mental states but rather the physical comfort of a heavily pregnant woman walking on ice. So long as Masha is not terribly in pain and can get home all right, the Good Samaritan is justified in judging her to be well. In the second case, the caring friend's concern is a richer notion of well-being—probably closest to what philosophers call 'well-being'. When she asks Masha how she is doing, she has in mind whether Masha is fulfilling her hopes and whether she is depressed. So the friend justifiably concludes that Masha is not doing well. Finally, the social worker is employing yet another notion of well-being, most akin to quality of life. To use Thomas Piketty's term, Masha is a member of the *patrimonial middle class*, which makes it likely that she will not fall through the cracks when she hits a vulnerable stage in life. She also has lots of people and resources to count on. Those two things are enough for well-being as far as the social worker is concerned.

Notice that both the threshold that separates well-being from ill-being—that is, how much of a given good Masha must have in order to qualify as doing well—and the factors that count for well-being appear to vary across our three cases. I call these *threshold* and *constitutive* dependence, respectively. How should philosophy accommodate them?

1.2. AGAINST CIRCUMSCRIPTION

The Circumscriptionist, recall, will deny that Masha's well-being is in fact in question in each of the three contexts. Instead, perhaps comfort and basic physical ability are in question for the Good Samaritan, and the quality of life of new parents is in question for the social worker. Only in the second context, namely the conversation with a close friend, are we properly, or perhaps nearly properly, talking about Masha's well-being. Circumscription is the modus operandi in philosophy. For example, a well-being hedonist treats well-being as the overall ratio of all positive over all negative mental states of a person's life or at a time. For other contexts, philosophers have drawn distinctions between many different notions all related to, but distinct from, well-being: welfare, quality of life, experiential quality, happiness, to mention a few. So, the objection goes, it is not the case that the concept of well-being exhibits the variability in question. Rather, there are many different concepts, each bearing some connection to well-being without being it.

The first problem with Circumscription is the lack of empirical fit with widespread and not obviously controversial linguistic practices. In everyday life, 'How are you doing? or 'How is it going?' are questions asked in a variety of contexts. And a variety of answers, holding fixed the state of the subject's life, are understood as appropriate replies. When asked if we are doing well, we do not necessarily start analysing whether our life as a whole instantiates values we endorse (if that is one's favourite theory of well-being). Instead we quickly figure out what the question means and adjust accordingly. I already granted that there are contexts when 'How are you?' is a greeting or a chat-up line.

But so long as it is a question about well-being in *more than one* context, the Circumscriptionist must face the lack of fit between this view and regular usage.

This problem is magnified when we turn our attention to the scientific, rather than the everyday, context. Here, we do not even need to assume that 'How are you doing?' is a question about well-being. Researchers all across the social and medical sciences use the term 'well-being' freely and abundantly in exactly the way the Circumscriptionist takes to be illegitimate. It is a term used to refer to a minimal quality of life in development economics; to a health-related quality of life in medicine; to a child's access to decent schooling, healthcare, and parental love in disciplines that study children; to mental health in psychiatry and clinical psychology; and so on and so forth. Economist Angus Deaton said in his 2016 Nobel prize lecture:

> The work cited by the Nobel committee spans many years, covers areas of economics that are not always grouped together, and involves many different collaborators. Yet, like the committee, I believe that the work has an underlying unity. It concerns well-being, what was once called welfare, and uses market and survey data to measure the behavior of individuals and groups and to make inferences about well-being. (Deaton, 2016, p. 1221)

Such examples are plentiful. The historical and social reasons for this are undoubtedly complex, but the fact remains that scientists are not shy about saying they are studying well-being.

A perfect fit with usage (should that even be possible) is, of course, not the sole criterion on which to judge a philosophical theory. Still, some such fit is desirable even for those who do not endorse the aim of conceptual analysis.[2] If the discrepancy between actual usage and the theory is substantial, then some explanation is required. This explanation should take the form of an error theory specifying reasons why, in practical contexts, both everyday and scientific, so many competent

---

2. See Haybron (2008, Chapter 3) for an articulation of a methodology in moral psychology that is constrained but not determined by linguistic usage.

speakers use 'well-being' and its cognates the wrong way. Perhaps a Circumscriptionist can provide such a theory. But there is something dogmatic about stipulating that so many uses of 'well-being' in science and everyday life are wrong.

This restriction also seems unnecessarily for normative reasons. It is fine to claim that the all-things-considered sense of well-being is primary and the most important, but the Circumscriptionist I set up makes a further claim that there is not any other kind of well-being evaluation. Some notion of well-being is arguably what is at work when we are called to make decisions that pit one value against another, say, time with children versus a higher salary.[3] But these decisions do not always consider all values at once. Sometimes we intend *some-things-considered* evaluations. For these cases there are theoretical costs to denying that well-being is the subject of Masha's three exchanges. Drawing distinctions between well-being and other things is a valuable skill that philosophers have perfected. But the more distinctions there are between well-being on the one hand and flourishing, welfare, quality of life, happiness, comfort, and so forth on the other, the narrower the scope of well-being becomes. So other options are worth considering.

1.3. THE DIFFERENTIAL REALISATION VIEW

In addition to Circumscription, a second dogma of philosophy of well-being is that the state of well-being is realised always by the same state provided it is characterised in a sufficiently abstract way. This way may be flourishing, or a good balance of pleasures over pains, or the fulfilment of a rational life plan, depending on which theory our Circumscriptionist endorses.

The Differential Realisation view, on the contrary, retains the semantic stability of 'well-being' but accommodates instances of threshold and constitutive dependence by instead allowing that well-being can be realised by different states of the world in different circumstances. This

---

3. Crisp (2013) makes this point against proposals to eliminate the concept of well-being altogether.

allowance is not just the uncontroversial liberality—different rational plans are appropriate for different people—but a stronger claim, for example, that well-being is sometimes fulfilment of a rational life plan and sometimes happiness. This variety is fully compatible with the existence of a single concept of well-being, a concept whose semantic content does not depend on circumstances.

Consider a parallel. Epistemologists have recently been working out the relation between knowledge and practical interests. As John Greco puts it: 'if the function of knowledge is to serve practical reasoning, it should be tied to the interests and purposes that are relevant to the practical reasoner at issue' (Greco, 2008, p. 433). But how exactly should it be so tied? Currently on offer are several options, but I consider the two main ones: attributor contextualism and subject-sensitive invariantism. The latter view, proposed recently by John Hawthorne (2004) and Jason Stanley (2005), with variations by Berit Brogaard (2008), has many affinities with what I call the Differential Realisation view of well-being. On this view, what varies is not the content but rather the truth conditions that realise the invariant semantic relation of 'know'. Part of these truth conditions can be nonepistemic facts about how much rides on the truth of a knowledge assertion, the standards and interests of the conversationalists, and even the speaker's personality.

Applied to well-being, the Hawthorne/Stanley/Brogaard view would state that the Good Samaritan, the friend, and the social worker are all making the same claim when they say that Masha is doing well or not well. However, the state of Masha's life that makes the sentence 'Masha is doing well' true shifts with circumstances: in the case of the Good Samaritan it is different than in the case of the social worker, for example. Included in these circumstances are nonprudential facts, so to speak, such as the relationship between the subject and the evaluator, whom the subject is being compared to, what the purposes of the evaluator are, and so on.

Richard Kraut appears to endorse semantic stability of well-being claims:

> the word 'good' does not vary its meaning as it is applied to these
> many diverse subjects [plants, animals, humans, etc.]. We say,

"That kind of barley is good for horses but not for human beings."
But we do not mean that it is good-in-a-horsey-sense for horses
and bad-in-a-human-sense for humans. (Kraut, 2007, p. 3)

For Kraut, the talk of goodness of a good G for a living thing S refers
to 'the conformability or suitability of G to S. It indicates that G is
well suited to S and that G serves S well' (Kraut, 2007, p. 94). But
I would not call Kraut a Differential Realisation theorist. The substantive theory of human well-being he endorses is a view he calls
developmentalism. According to it well-being is flourishing, that is,
the 'maturation and exercise of certain cognitive, social, affective, and
physical skills' (Kraut, 2007, p. 141). (We encounter this proposal in
Chapter 3 on child well-being and then again in Chapter 5 on measurability.) Naturally, the actual state that realises flourishing depends
on what kind of living being one is (e.g., whether one is a toddler, or
an adult human, or a horse) and on the environment this being is in.
But this is a rather limited degree of differential realisation. So in my
picture Kraut's view is Circumscriptionist. Differential Realisation,
on the other hand, takes a further and more controversial step to allow
that walking safely on ice, or having all a new mother needs, realises
Masha's well-being in some situations.

Differential Realisation could endorse a full-blown pluralism about
the existing Big Three theories or find more modest versions to accommodate threshold and constitutive dependence. For instance, a hedonist
might say that a certain quantity of positive mental states will realise
well-being in one set of circumstances but not in another. Or, depending on whether we are talking about the context of medical treatment
or a music performance, different kinds of positive mental states will
qualify as well-being. Similarly, a desire-based theorist could specify
that satisfaction of different kinds of desires, or their satisfaction to a
different degree, could realise well-being in different contexts. In each of
these cases the theorist would maintain that well-being assertions carry
the same stable meaning across different situations but have context-
sensitive truth-makers.

What is this context-invariant meaning of 'good for' that the
Circumscriptionist and the Differential Realisation theorist both

assume? Kraut's (2007) suitability analysis is only one; I discuss others in Section 1.7. But before that, let us consider the third option.

1.4. CONTEXTUALISM ABOUT WELL-BEING

Whether an expression's content is sensitive to context is a question about whether this content is, at least in part, indexed to a context, which is why some content-sensitive expressions are called indexicals.[4] Classic indexicals, such as 'I', 'here', and 'now' are thought to have more content than is just expressed in the corresponding words. 'I' refers to me, Anna, when spoken by me; 'here' refers to my hometown while I am there, and so on. John Perry (1998) calls this extra content the 'unarticulated constituent' of a context sensitive proposition.

There is a debate in philosophy of language about whether this unarticulated constituent really is part of the semantic content of a proposition or merely conversational pragmatics (Cappelen & Lepore, 2007). Those who insist on understanding it as part of the semantics are known as contextualists, because they insist that the content of propositions changes depending on the context in which they are uttered. Indeed they insist that the proposition is not even there in the absence of the context. Radical contextualists claim that this is true of much of our language, not only classic indexicals. In this vein, recently contextualism has been extended to terms and expressions about knowledge. On this view the content of propositions that attribute knowledge varies with context. The context is defined not just by what evidence there is about the truth and justification of a belief but also by practical circumstances, for example, how much rides on the proposition being false. When not much rides on it, the standards are lower than otherwise and hence it takes less for a knowledge sentence to come out true. This is how the same sentence about the same subject with the same amount of evidence (say, 'Anna knows she has hands') can come out true in one context (say, an everyday conversation)

---

4. For related but distinct applications of contextualism to ethics see Unger (1995), Norcross (2005), Northcott (2015), Brogaard (2008). Tiberius (2007) and Tiberius and Plakias (2010) partially anticipate some elements of the view proposed here.

and false in another (say, a philosophy seminar). It happens because in fact different propositions are being asserted.[5]

Applied to well-being, contextualism would treat 'well-being' and related expressions as terms whose semantic content depends on the practical situation. Thus 'Masha is doing well' (or 'Masha is not doing well') express different propositions in the mouths of the Good Samaritan, the friend, and the social worker. What is this difference? Partly, it is the difference in the threshold. Just as the standards of justification vary between a philosophy seminar and an everyday conversation, so the threshold of well-being can also vary. Simply seeming to be all right to a stranger is all 'Masha is doing well' means to the Good Samaritan, but it means something more to Masha's friend or the social worker. Similarly, the threshold can shift depending on whom Masha is being compared to (mothers-to-be from Darfur or from Monaco). But the threshold is not the only thing that varies. It is not that the social worker has a lower standard of well-being than the best friend. On the contrary, she might have a higher one. Rather, the social worker is focusing on different things than the friend. Her project is to ensure that parents have all the basics and have someone to talk to regularly. It is irrelevant to her that they also realise their talents and deeply held dreams. According to contextualists her notion of well-being is qualitatively different. This is constitutive dependence.

Contextualism can vary in scope: a radical version could deny that there is any common content between different well-being assertions, while a less radical one could allow for some common core. But before we get into further detail, let us consider evidence for context dependence.

1.5. DOES 'WELL-BEING' PASS THE TESTS OF CONTEXT SENSITIVITY?

Some uncontroversial cases of context dependence are comparative adjectives such as 'tall', 'intelligent', and 'poor'. They are gradable in that 'tall' for a wrestler is not the same as 'tall' for a volleyball player.

5. See Rysiew (2007) for an overview of epistemic contextualism.

'Well-being' clearly admits of the sort of gradability a contextualist might appeal to. We talk of, say, a victim of a car accident who is recovering as 'doing well for someone in his situation', or of a person left homeless and destitute by hurricane Katrina as 'doing terribly for an American citizen'. Jason Stanley (2004) points to a lack of gradability of expressions about knowledge as evidence against epistemic contextualism. He proposes a two-stage test of gradability. An expression should allow for (a) modifiers such as 'very' or 'really' as in 'He is very or really tall' and (b) comparative constructions as in 'taller than'. He argues that expressions involving 'know' lack both of these features and therefore that the term 'know' is not gradable (Stanley, 2004). Expressions involving 'well-being', on the other hand, clearly do pass Stanley's test: 'doing very well' or 'doing better than someone' are acceptable even for those who do not believe in a single metric of well-being. So in this respect well-being contextualism is easier to defend than epistemic contextualism.

How else could we tell if the talk of well-being is context-sensitive? There are at least two tests widely discussed in the literature, due to Cappelen and Lepore (2005). The first test is Indirect Disquotational Reporting (IDR): context-sensitive expressions are more difficult to report indirectly disquotationally. 'I am here' does not easily get reported as 'Anna said I am here'. If IDR is easy, then the expression is most likely not context sensitive.

Take the expression 'Masha is doing well'. At first it seems easy to report it indirectly without the quotation marks: 'The Good Samaritan said Masha is doing well'. But now imagine two close friends of Masha talking about how Masha is doing in that significant tone of voice. The first one is worried about Masha due to her impending depression. The second one replies, 'Well, I know that a guy who helped her on the street thought she is doing fine', effectively reporting the Good Samaritan's judgement on Masha. There is something clearly wrong with the second friend's implication that the Good Samaritan's judgement is at all relevant to the issue the two friends are discussing. She is not reporting his judgement appropriately, for in the context discussed by Masha's friends a different sense of well-being is at play. The Good Samaritan clearly meant something else. If, on the other hand, another stranger on the street who also saw Masha slip and fall reports the Good Samaritan's judgement about Masha, the context is preserved and the indirect reporting without quotation marks works. So it seems that, at least sometimes, the talk of well-being does behave as IDR requires.

The second test of context-sensitivity is collectivity. Say two people A and B uttered a single sentence S. If we can collectively report these utterances as 'A and B both said that S', then S is probably not context-sensitive.

In our case this requires combining the reports about Masha of, say, the social worker and the Good Samaritan: 'The social worker and the Good Samaritan both said that Masha is doing well'. Does this work? Again, not really. If we remember that they uttered it in very different contexts employing entirely different standards, then the report does not look right.

In each case, IDR and collectivity, the claim that the talk of well-being passes the tests of context-sensitivity depends crucially on the denial that the all-things-considered sense of well-being is the only legitimate sense out there. If we insist that only the conversation with the friend counts as a conversation about well-being, then the judgements by the social worker and the Good Samaritan will be dismissed. So long as they are allowed— as I urged earlier in the discussion of Circumscription—contextualism about well-being can go through.

1.6. WHOSE CONTEXT?

The next crucial detail is the locus of context. Whose context gets to determine the truth value of a proposition? The main candidates are the subject, the speaker, and the evaluator. Epistemic contextualists, for example, largely agree that it is the interests, practical purposes, and expectations of the speaker that make the context what it is. Relativists prefer to focus on the context of the claim's evaluator and invariantists on that of the subject (hence the name 'subject-sensitive invariantists') (Rysiew, 2007).

My sense is that to explain instances of threshold and constitutive variance, well-being contextualism should be formulated by reference to the speaker's, not the subject's, context. This is because no significant changes take place in Masha's life as she comes across the Good Samaritan, the friend, and the social worker. It is because *their* contexts are so different that 'Masha is doing well' comes out to have different truth values. By the same token, it is the context of the development economist evaluating a poor community (rather than of his or her subjects) that makes it the case that minimal quality of life, and not some medical notion, is appropriate. Of course, in the case of self-evaluation the speaker is the subject.

However, there is more than one way of fleshing out a speaker's context. I follow John Hawthorne's (2004) distinction between *salience* and *practical environment*. Roughly, if context is understood in terms of the salience of certain possibilities of error to the speaker, then the truth of a knowledge claim will ride on the speaker's psychological state. If the possibility of a certain error is not salient to this person, then they will be entitled to treat the relevant proposition as knowledge. On the other hand, if the context is understood in terms of the objective practical environment of the speaker, then the possibility of error (i.e., the possibility that not-p) bears on knowledge of p regardless of whether this person happens to know of this possibility or be in a psychological state in which this possibility is salient.[6]

This is a helpful distinction for the case of well-being. A contextualist takes the semantic value of a well-being claim to depend on context, which in turn can be understood either as what is salient to the speaker or else as the objective features of the situation in which the speaker finds themselves, and that may or may not be salient to the speaker. By objective features, I mean those features that make it the case that a certain notion or standard of well-being applies, whether or not they fall within anyone's field of awareness. These features may well (and should) include the agent's values, commitments, and attitudes, but they are not exhausted by them. If Masha values intellectual life, then that can partly constitute her well-being just in virtue of her attitude. But she can also be wrong about whether or not a given standard of well-being applies in a given case. So although objectivity here does not ignore the agent's point of view, it does mean that the agent's actual

6. The relevant sections of Hawthorne (2004) are 4.2 to 4.3.

judgement about what standard of well-being is locally appropriate is not the final authority.[7]

A purely psychological understanding of context, as *just* the features that the speaker finds salient (either as constitutive of well-being or as being the right threshold), leads to ugly consequences. Suppose Masha's friend adopts a dark minimalist view of well-being as mere survival. Then she might well take Masha's impending depression to be irrelevant to her well-being, as paying attention to emotions is mere effeteness. She might judge Masha to be doing perfectly well despite her malaise. This is an unwelcome consequence. I would prefer a contextualism that rejects the purely psychological interpretation of context. Whether or not a certain notion and threshold of well-being applies in a given situation may or may not be known and/or accepted by Masha's potential benefactors, and it is a big question what makes a sense of well-being appropriate in a given case.[8] But for now I only want to formulate a plausible version of well-being contextualism.

To summarise, such a view treats well-being expressions as having context-sensitive semantic content, at least to some extent, where context is fixed by the features of the practical environment of the speaker at the time in which the judgement of well-being is made. Which features of the environment? Generalising from examples we have encountered so far, constitutive variation depends on the nature of the relationship between the speaker and the subject (the Good Samaritan is a stranger, hence his minimalism about Masha's well-being, whereas the friend has

---

7. Earlier I mentioned the well-known but controversial results in psychology that appear to show a dependence of judgements of life satisfaction on the context of evaluation (Strack et al., 1990). In these experiments, the presence of a person in a wheelchair had the effect of increasing self-reports of life satisfaction. (See Chapter 1, footnote 12 on replicability of these results). This is plausible evidence that people adjust their standards of well-being depending on context, a welcome datum for a contextualist (though not only for them). I take these results to be relevant but not decisive. Contextualism about well-being does not need to accept uncritically the actual usage of well-being terms, which is why I reject a purely psychological interpretation of context. And even sticking to actual usage as the main evidence, we often criticise people's assessments of their own well-being. This possibility must be preserved by any plausible version of contextualism.

8. Of course, the same qualifications can be applied to the Differential Realisation view. Which truth-makers realise well-being in different contexts does not need to be down to anyone's whim.

different obligations). The threshold variation depends on the relative badness of the subject's situation (Masha has fallen on the ice, hence the Good Samaritan's concern).

Now we are in a position to evaluate the relative advantages of Differential Realisation and Contextualism.

1.7. IS CONTEXTUALISM PREFERABLE TO DIFFERENTIAL REALISATION?

Before examining the relative advantages of the two main options, it is worth voicing a possibility that the sort of variability 'well-being' exhibits is due to the very meaning of the term, not due to the changes in extra-linguistic context. 'Well-being' might be polysemous; that is, it might have several distinct but related meanings. If it was, that would be a fourth option to consider. 'Healthy', 'mole', 'get' are likely polysemous to the extent that 'healthy attitude', 'garden mole', and 'I get you' mean something distinct and yet similar to, respectively, 'healthy society', 'mole in an organisation', and 'I get excited'. If 'well-being' is polysemous, then it is no wonder we find threshold and constitutive variation and a variety of constructs called 'well-being' across different sciences. Testing for polysemy would take me too far afield. My bet is that even if well-being is polysemous, it is not *only* that. Polysemous terms often translate into different terms in different languages, which appears to be the case for 'well-being'. In Russian, for instance, it can translate as 'blagopoluchie' or, closer to 'welfare', 'blagosostoyanie'. But within 'blagopoluchie', there is still also space for context-induced variation in meaning. Thus, 'blagopoluchie' when speaking of a family may not mean the same as when speaking of an employee, or a patient, or a community. It is thus fair to continue exploring Contextualism and Differential Realisation.

Both views take seriously the diversity of notions and constructs of well-being in everyday life and the sciences, and thus both score major points against the Circumscription view. Beyond that, what reasons are there to prefer one over the other? Two advantages of Contextualism over Differential Realisation make it prima facie preferable. The first is that Contextualism avoids the counterintuitive consequence that well-being can come and go with changes in context. On Differential Realisation Masha can actually improve her well-being by moving from one context to another. The second problem with this view is its need to postulate an invariant semantic content of well-being assertions. I discuss these two problems in turn.

First, Contextualism does not imply that well-being, or, more precisely, the family of properties picked out by this term, depends on context. As many epistemic contextualists have been careful to point out, what varies with context is not whether one knows but only whether one counts as knowing (DeRose [2000] and many others). This is why it is consistent with epistemic contextualism to claim that whether or not one knows (in all the meanings of 'knows') does not depend on context. Indeed, most epistemic contextualists take this option, for they do not want their view to imply that knowledge may come and go in a single conversation or with a change in perspective (DeRose, 2000). Similarly for well-being contextualism: the particular well-being phenomenon picked out by a context does not itself depend on context. Thus it is not the case that Masha was doing fine when talking to the Good Samaritan, then suddenly started doing badly in the company of her friend, and then again started doing much better when the social worker showed up. That would be the case if we had adopted the Hawthorne/Stanley/Brogaard view, according to which knowledge (or, in our case, well-being) actually comes and goes as the practical situation changes. The counterintuitive nature of such coming and going is a prima facie reason against Differential Realisation and in favour of Contextualism. Differential Realisation implies that Masha can actually improve her well-being just by changing the context of evaluation. The secret of a good life would then be to always place oneself in the least demanding context!

Of course, this is just one data point. Speaking of the knowledge case, Stanley (2007) points out that one counterintuitive feature alone cannot be the deciding objection since contextualism is not devoid of its own weird consequences. But we have already seen gradability of well-being terms, with respect to which well-being contextualism is better off than epistemic contextualism. So Differential Realisation about well-being does face a disadvantage.

The second challenge is whether this view can specify the invariant semantic content of well-being assertions. If 'Masha is doing well' expresses the same proposition in any context in which it is uttered, what does well-being mean?

Stephen Campbell (2013) distinguishes between four possibilities:

(a) Well-being assertions make a claim about what is suitable for a being or what serves it well (Kraut, 2007).

(b) Well-being assertions make a claim about what one should wish for people insofar as one cares for them (Darwall, 2002).

(c) There is also *locative* analysis: well-being is about being good and being located in the life of the being in question.

(d) And finally, there is Campbell's own *positional* analysis: well-being of a person is the desirability of this person's position.

Elsewhere Campbell (2015) argues that these concepts are also conflicting, not just plural. This is, of course, grist for the contextualist's mill. But even if we do settle on a single such concept, context dependence will still reappear.

Darwall (2002, Chapter 4) for instance, takes as primitive the notion of care or sympathetic concern—an attitude one develops toward a person one values for their own sake. Care or sympathetic concern amounts to a different attitude or emotion depending on the situation, and justifiably so. To be concerned for one's own child is one attitude, for a homeless person on the street on a winter night another, for civilians in the war zones yet another, and so on. Depending on one's relationship to the object of care, on the severity of this person's situation, and perhaps other factors, care is a concern about whether or not the person is reaching his or her full potential, whether or not this person will freeze on a cold night, whether or not the person will survive the war, and so on. These differences are both psychological and normative. It is appropriate *not* to be concerned that a person in a war zone comes into contact and appreciates the aesthetic value of Beethoven's late string quartets, whereas it might be appropriate to be concerned that one's child or partner does. Adjusting care is not abandoning it. Rather 'care' itself is not univocal. And with context sensitivity of care comes context sensitivity of well-being claims, since on Darwall's view care and well-being are so tightly connected.[9]

It is thus natural to think of Darwall's (2002) proposal as one way of specifying the common conceptual core of all well-being assertions: to do well is to have what the rational carer would want for you (option [b] above). But note that even if we accept Darwall's story, it does not tell us the full semantic content of well-being claims but only the structural core of this content.[10] This core is filled out differently depending on circumstances. Thus when the Good Samaritan, in his role as a rational carer, inquires about Masha's well-being, his question has a very specific and locally appropriate content—'to do well' in this context means 'to have what a caring rational stranger on the street would want for a heavily pregnant woman walking on ice'. If so, claims about well-being have a complex structure: along with the common structural core (which Darwall's account may help to identify), they also carry a more specific semantic content—*that which the rational carer in that particular situation would want for the person in that particular context*. The second and third 'that' are clearly context-sensitive, which means 'well-being' will end up being context sensitive too. Darwall's proposal is thus easily co-opted by Contextualism for its own purposes. But its original version *cannot* fully specify the context-invariant semantic content of well-being assertions that the Differential Realisation theorist needs.

Proposal (c)—well-being as a claim about suitability—faces an analogous obstacle. What is suitable for Masha or what serves her well is best

---

9. I thank Dan Haybron for this idea. One might object that what are sensitive to context are the duties that the carer has toward their subject, not what caring is. To be sure, duties do depend on context, but it is also plausible to think that the content of the attitude of caring also varies with context.

10. Kaplan's (1989) notion of character meaning as opposed to content can be used to characterise Darwall's or any other such proposal. The classic indexical 'here' has one character meaning but many different contents. Similarly, a proposal for the conceptual content of 'well-being', such as Darwall's, may be able to specify its character meaning but not its content.

thought of as specifying the core meaning of the many assertions about Masha's well-being, but it does not exhaust the content of these assertions. The content has to make mention of suitability *given the context.*

We see here a pattern: proposals on the nature of the concept are more plausible for identifying the common core meaning of these assertions but not their full semantic content. Plausibly, this content varies with context just as the contextualist claims. It is thus hard to pinpoint the invariant content of well-being assertions that the proponent of Differential Realisation requires.

Note that the contextualist can feel free to adopt any one or more of the four proposals for their own purposes, though I do not have a strong view on which one. The contextualist just does not think that these proposals exhaust the content of well-being assertions.

This is not a full score card of the relative strengths of Differential Realisation and Contextualism, but I submit it is a sufficient first pass to treat Contextualism about well-being as a frontrunner.

1.8. WELL-BEING CONTEXTUALISM'S DOS AND DON'TS

To summarise, Contextualism is a view that well-being expressions have varying content depending on the context in which well-being is assessed. This context is fixed by the features of the practical environment of the speaker at the time when the judgement is made. These features can include facts about the subject's values and commitments, the relevant contrast classes (e.g., to whom the subject is being compared), what the relationship between the subject and the speaker is, what resources are available to the speaker qua potential benefactor, and perhaps many others.

Is Contextualism controversial? I think not for four reasons. First, contextualists need not be eliminativists about well-being. They need not deny that there is a common core meaning to all well-being assertions.

Contextualism does not, by itself, speak against the possibility of a general substantive theory of well-being. Scepticism about such a theory animates the next chapter—perhaps one all-purpose theory of prudential good is a philosopher's dream. But such worries are orthogonal to the issue at hand here. Contextualists need not deny the value of the traditional philosophical all-things-considered approach to well-being. As well as telling us what well-being amounts to in contexts calling for such a comprehensive evaluation (e.g., Masha's conversation with her close friend), the traditional approach may also be able to specify the conditions under which different notions of well-being are appropriate. All that Contextualism requires is the abandonment of the assumption that 'well-being' and its cognates have a stable and narrow semantic content.

Second, Contextualism need not amount to a promiscuous 'let all the flowers bloom' or 'anything goes' view. It is not committed to every conceivable concept of well-being being applicable in some context. It may be that there is a context suitable to each concept of well-being or that some concepts apply in more than one context, or that some concepts apply in no context at all. Contextualism merely seeks a mapping from contexts to concepts of well-being. By the same token Contextualism does not imply an arbitrary proliferation of well-being concepts. It does not commit us to the existence of 'left foot in November well-being', 'waiting in line at the post office well-being', and other absurdly fine-grained and trivial versions. Assessments of well-being typically enable us to systematise our knowledge about the state of a person and to deliberate about what we ought to do for that person. So individuation of contexts must not be wanton. If an assessment of 'left foot in November well-being' really did play an important role, then it would qualify as a legitimate context-specific notion well-being; if not, then it does not.

Third, Contextualism does not imply anarchy and miscommunication. We already use many different notions of well-being and manage to communicate just fine without talking past each other. A friend asks me 'How is your grandfather doing?' talking about an 80-year-old with advanced Parkinson's. I answer her question by describing his level of physical comfort and his emotional state, keeping in mind what is and is not reasonable to expect in this situation. We then start talking about how I am doing and I tell her about my young children, my family across continents, my stresses and anxieties. Our communication is not impeded by the subtle shift in context. Something like this happens in science and policymaking—well-being standards shift often without explicit acknowledgement. Of course, it would be best if such shifts were explicit and publicly justified, and Contextualism would naturally enable such a conversation. Once the contextualist semantics of well-being expressions are accepted, we can go ahead and build an explanation of which thresholds and notions of well-being are appropriate in which contexts.

Fourth, Contextualism does not disallow comparisons of well-being of people in very different situations. Masha can be compared with other people and with other time slices of herself. We just need to be clear about the context of this comparison to ensure that we are using the same notion of well-being all the way through. We cannot just compare the Good Samaritan's judgement of Masha's well-being to that of the social worker's. Comparisons necessitate a single context. Sometimes this notion will be the philosopher's favourite all-things-considered notion, and sometimes it will not.

Adopting contextualism about well-being is the first step toward evaluating the appropriateness of constructs of well-being and for judging the practical relevance of particular scientific findings about it. That said, the exact scope of Contextualism is up for grabs—that is, how radically the content varies and how many different notions of well-being there are. It seems plausible that Contextualism explains some variation and hence is a partial explanation for construct pluralism. But it is not a full explanation. Some variation in scientific constructs is contextual—health-related quality of life is probably just a different notion from national well-being. But some differences likely indicate deeper differences. To those I now turn.

In *Family Happiness* Tolstoy is interested in what I have called here one of the many contextual notions of well-being, which in his case is, well . . . family happiness—a good life in a union with another person. To understand how, as Masha discovers, well-being can take such different forms throughout life we need a substantive theory of well-being, which will be the focus of the next chapter.

Chapter 2

Is There a Single Theory
of Well-Being?

The previous chapter put forward contextualism about well-being—roughly, the view that there are several notions legitimately referred to as 'well-being' that apply in different circumstances. Contextualism goes some way toward explaining why, in life as well as in science, so many different notions are employed. But it does not tell us what the notions refer to and what justifies them. It tells us that the Good Samaritan, the social worker, and the best friend all mean something different when they speak of Masha's well-being, but it does not tell us what Masha's well-being in each of those contexts properly amounts to. Is the Good Samaritan right to focus only on Masha's ability to get home all right? Is the social worker right to disregard Masha's anxiety and dissatisfaction? Is the friend right to expect that Masha should lead a life that best realises her dreams? These are all substantive questions about the nature of well-being, and they cannot be settled just by observing language.

To tackle these questions we need to know what well-being (or well-beings) actually is. Traditionally this knowledge is enshrined in theories that were a mainstay of ancient philosophy and remain so today. Nothing in Contextualism denies the existence of a theory of well-being. There may be many concepts, but which state of the world each concept refers to may still be decided by appeal to a single correct theory, any of the Big Three—hedonism, subjectivism, or eudaimonism—or some improved hybrid view.

In this chapter I discuss the possibility of such a theory in the context of the needs of the science of well-being. Do we have such a theory? Can it do the work that we need it to do? What should we expect from theories of well-being more generally?

Appendix A is a review of the state of the art in the analytic philosophy of well-being. Readers unfamiliar with this material should consult it first. Here I go straight into those of its features that bear on the questions of this chapter. My argument is as follows: Observe how philosophers deal with problems their theories face. Such problems are typically intuitive counterexamples—imaginary but plausible scenarios in which there is well-being but the theory is not satisfied or in which the theory is satisfied but there is no well-being. They force a theory's advocate either to bite the bullet or to make the theory more intricate, the latter being the much more common move. But greater intricacy, though it makes for a more defensible theory by philosophers' standards, typically compromises the connection between theory and measures of well-being. When philosophical accounts are used by scientists, they are used as *models* rather than as theories. A model, in this sense, is a conceptual tool for building a measurement procedure. Unlike a theory, which fully specifies how it should be used, a model requires additional outside knowledge. Once we see that the science of well-being treats philosophical proposals as models, it is natural to think that there are many such models and that there is no single overarching model to regulate their use.

I call this view *variantism* about well-being. *Invariantism*—the view that there *is* a single theory of well-being that underpins the variety of constructs of well-being in life and science—has been the modus operandi in philosophy. Is it a good one? Its advocates will no doubt appeal to the importance of starting with an assumption that knowledge about well-being can be unified, and I grant that the case can go either way. But I still wish to put variantism on the intellectual map and to give reasons to take this view seriously, if only because formulating it yields a more realistic view about what we can expect from a theory of well-being and what theories we are better off pursuing. Just as we have been contextualists all this time, we may well have been variantists too.

2.1. HOW PHILOSOPHERS REACT TO COUNTEREXAMPLES

Theories of well-being in today's philosophy, the Big Three, are theories about what is noninstrumentally good for a person or a being. They are not primarily theories about how to lead a good life, how to be happy, or how to find meaning in life. In fact they are not 'how to' theories at all. Rather they seek to characterise well-being at the most general level, leaving it for others to say how to pursue it in a given situation. Typically well-being is distinguished from happiness on the one hand (a psychological state) and choiceworthiness on the other (the property of being supported by the weight of reasons). In the contemporary Anglophone tradition, theories of well-being are usually put forward by philosophers broadly interested in the nature and character of values and hence have typically arisen as part of ethics, meta-ethics, and moral and political philosophy.

Given the number and intricacies of these theories, classifying them into groups is a research project in its own right. Derek Parfit's (1984) classic partitioning distinguishes between mental state, desire fulfilment, and objective list theories—hence my choice to speak of the Big Three. I am not particularly wedded to this classification, nor do I have my own theory of well-being to contribute. Rather, I am interested in what happens when a given theory of well-being faces objections of the classical sort:

- Hedonism is wrong because things other than mental states matter.
- Desire fulfilment theories are wrong because people can desire what is bad for them.
- Objective list theories are wrong because a person may not benefit from a given good.

This is not a comprehensive list, but it will do. Much effort in philosophy goes into pondering the implications of these claims and into formulating replies to fix the alleged problems. I now turn to these fixes with an eye on exposing their heavy costs for the theories' usability.

How do philosophers respond to counterexamples? They commonly adopt one of the following strategies: *bite the bullet, go hybrid, go intricate*, or some combination of these. Biting the bullet involves admitting that an inconvenient scenario is indeed allowed by the theory but maintaining that the theory is still preferable to others—perhaps because it is a better overall package or because the inconvenience of the scenario is an illusion. By contrast, going intricate requires amending the theory with a new element or a new constraint. It sometimes amounts to going hybrid: if the new element is really part of another theory of well-being, then the resulting account is a hybrid account. Let us look at some examples.

A good example of bullet-biting is provided by Roger Crisp's (2008) defence of hedonism. For Crisp, no decent version of hedonism can deny that the denizen of a Nozickian experience machine is doing as well as the person with identical experiences who lives their life for real. But the intuition that the latter life is preferable to the former on prudential grounds is, for Crisp, an illusion that has an evolutionary explanation. To value achievement, authenticity, and whatever else the experience machine cannot give, on their own and independently of experiences they cause, is 'a kind of collective bad faith' that results from living in small groups whose livelihood crucially depends on its members' achievements in hunting or gathering (Crisp, 2008, p. 639).

An example of bullet-biting on the part of a desire fulfilment theorist is to say that the fulfilment of any—even trivial, meaningless, or evil—desires contributes to well-being or on the part of an objective list theorist to say that the failure to enjoy or desire an objectively valuable good does not diminish the contribution of this good to one's well-being.[1] Bullet-biting creates the demand to explain away the problematic intuitions or to showcase compensating advantages of the theory whose advocate bites the bullet. Philosophers typically appeal to the advantages of the package as a whole even if some element in it is counterintuitive. But in itself bullet-biting does not undermine a theory's ability to be used in science.

---

1. Keller (2004) bites the bullet on behalf of desire fulfillment, Arneson (1999) on behalf of objective list.

Thus thanks to Crisp's (2008) explaining away of the experience-machine intuition, scientists with hedonist measures of well-being receive a license to ignore whether their subjects are in an experience-machine or otherwise radically deceived.

Things are different usability-wise when philosophers choose to go hybrid instead of, or in addition to, biting bullets. Perhaps the classic example is John Stuart Mill's version of hedonism, which incorporates the distinction between higher and lower pleasures. The higher pleasures cannot be traded off against any amount of lower pleasures because the higher pleasures are more noble and better suited to rational humans. This move sometimes leads commentators to classify Mill's theory of well-being as closer to Aristotle than Bentham in that it relies on a robust conception of human dignity, or at least as a hybrid combining elements of classical hedonism and eudaimonism (Brink, 2013).

There are many other hybrids around. Parfit (1984, p. 502) suggested amending objective list theories with the requirement that the list also be endorsed by the agent. Feldman (2004, pp. 109–114) proposed a reality constraint on hedonism such that only certain pleasures contribute to well-being. Haybron's (2008, Chapter 9) theory of well-being as self-fulfilment, defines this state as encompassing both emotional fulfilment (which for him is happiness) and a life in which one's values are realised. In each of these cases the hope is to unify in one account several intuitively valuable characteristics: happiness, the endorsement of one's projects, the objective goodness of those projects. Typically the unified elements become individually necessary and jointly sufficient for well-being. They are not to be traded off against each other or treated as individually sufficient.

Sometimes philosophers respond to counterexamples by going intricate rather than hybrid. Going intricate adds a modification to a theory without making it hybrid. When a desire-based theorist insists, as most typically do, that the desires whose fulfilment counts for well-being must be informed and rational, they add an extra condition to the original simple view that satisfaction of any desire is well-being constitutive. This is a more intricate version of the desire theory but not a hybrid version. The information and rationality requirements have a motivation independent of other theories of well-being. We want our desires to reflect reality because then these desires suit us better than otherwise—at least this is how desire theorists mean this amendment to work (Sobel, 2009).

Another example of the 'go intricate' move is Roger Crisp's (2006) modification of hedonism to accommodate the value of higher pleasures. Mill took seriously the objection that classical hedonism was a Philosophy of Swine—it allows lower pleasures, if sufficiently long and intense, to outrank the nobler higher pleasures in value. This is why Mill, on some interpretations, went hybrid and grounded the value of higher pleasures in human nature. To avoid going hybrid in the same way, Crisp identifies enjoyment not with the actual token experiences but with the common feature of all enjoyable experiences. Experiences have intensity and duration, but enjoyment does not play by the same rules. The value of enjoyment is in its phenomenological quality, which may well ignore the intensity and duration of the experiences. I may enjoy a conversation with a friend for its glorious intimacy, understanding, and laughter, and this judgement is mine and mine alone. It may well deviate from the score this experience receives on the simple intensity/duration scale. It deviates just because I enjoy it more, not because it is more noble or appropriate to humans or more desired. Hedonism is thus preserved.

Examples of both 'go intricate' and 'go hybrid' moves abound in philosophy.[2] They are not the only options available. One can also restrict the scope of the theory such that it would only apply in certain domains and not universally. But it is hard to think of any philosopher who takes this option. It is just not the done thing because the goal in the words of one of the contributors is 'to capture the whole truth about well-being' (Sarch, 2012, p. 441).

But such ambition is costly. For when theories become more intricate or hybrid, invariably they sacrifice their connection to measurement.

---

2. Jason Raibley's (2010) agential flourishing view and Dale Dorsey's (2012) belief subjectivism are arguably intricate versions of subjectivism.

2.2. THE COSTS OF THE STANDARD METHODOLOGY

In the case of intricate hedonism, in which the value of enjoyment is independent of intensity and duration, Crisp himself freely admits this consequence:

> It may have been a dream of some hedonists—Bentham perhaps—that one could invent some kind of objective scale for measuring the enjoyableness and hence the value of certain experiences, independently of the view of the subjects. But that—as Mill and Plato saw—is merely a dream. (Crisp, 2006, p. 633)

The scientists however keep on dreaming. Currently the best method for measuring affective states is previously mentioned experience sampling, famously adapted by Daniel Kahneman and his colleagues to happiness. He proposes a notion of 'objective happiness' measured by the temporal integral of 'instant utility'. Instant utility is how a particular moment feels to an individual on the good/bad dimension. It is derived from reports of emotions in the current moment and aggregated into a ratio of positive to negative emotions. Such instant ratings can then be integrated into 'total utility', approximately the product of average instant utility and duration. Then these ratings are combined to form the subject's 'hedonic profile'. If the horizontal axis represents duration and the vertical axis represents the level of instant utility, then the area under the curve refers to what Kahneman (1999) once called 'objective happiness'. It is objective in the sense that the subjects themselves do not judge their overall happiness but only their happiness at a given moment. This sense of 'objective' is not to be confused with objective theories of well-being, where objectivity has to do with goods that are good for an individual irrespective of their desires or attitudes.

Such aggregation only works if the following assumptions hold:

- Instant ratings must contain all the relevant information required.

- The scale has a 'stable and distinctive' zero-point, roughly equivalent to a 'neither good nor bad' attitude.
- The measurement of deviations from zero are ordinal rather than cardinal.
- A single value can summarise the good/bad evaluation.
- Such conscious or unconscious evaluation is constantly going on in the brain.

If these conditions hold, and Kahneman argues that they generally do, then the total utility refers to 'objective happiness'.[3]

Whether it should indeed be thought of as objective and as happiness is not the question here. Rather, I want to observe that Kahneman's method does not measure Crisp's enjoyment. Instead it measures, as well as we could hope at the moment, the standard Benthamite balance of positive to negative experiences, a characterisation Kahneman himself gladly accepts. Now that Crisp has redefined hedonism in terms of enjoyment rather than individual experiences and thus has made it immune to the sort of aggregation that allowed the Philosophy of Swine objection, it can no longer be measured by the best current procedure for measuring affective states. Kahneman's first assumption, and possibly others too, is violated. Hedonism remains an inspiration for this scientific project, but the scientists no longer follow the letter of the strongest version by the lights of philosophers.

Other instances of the 'go intricate' and 'go hybrid' moves regularly follow a similar pattern: an original philosophical proposal inspires a scientific measurement procedure, but later more defensible versions of this proposal do not get implemented in measurement. Mill's hybrid theory requires separate scales for high and low pleasures, whose integration is either impossible or very tricky. No one I am aware of has attempted it. Fully informed desires are, naturally, impossible to measure even indirectly. They are dramatic idealisations, supposedly in the same way that frictionless planes are. Creative social scientists go to

---

3. The theory behind 'objective happiness' is worked out in (Kahneman, 1999; Kahneman et al., 1997). Kahneman and Krueger (2006) and Kahneman et al. (2004b) propose U-index which is a macro-indicator of 'objective happiness' in a population.

great lengths to identify and measure preferences that best reflect individuals' considered judgement, or just plain values that people hold.[4] But none of these valiant efforts come anywhere close to the high standards of well-being typically formulated in the intricate or hybrid theories of subjectivist philosophers.

Even when the terms are the same, the concepts are anything but. Wayne Sumner's (1996) theory of well-being is authentic satisfaction with life. But whether this is measured by the Satisfaction with Life Scale depends crucially on just how strictly we interpret authenticity. And given the disciplinary norms of philosophy—namely, generality and exceptionlessness—the temptation is away from measurability.

Hybrid theories, before they are even considered as candidates for measurement, need to specify the precise level of each component good required for well-being and whether any of these goods will trade-off against each other.[5] Philosophers who debate the merits of objective list versus other views on intuitive grounds have yet to consider these issues, so central for measurement and yet so marginal given the disciplinary incentives of philosophers. Sometimes in response to counterexamples philosophers even advocate abandoning the assumption that any intrinsically valuable good enhances well-being always and everywhere irrespective of context, arguing instead that goods only have a disposition to do so, which may or may not be actualised (Fletcher, 2009). Once this step is taken, a philosophical account is perhaps less open to counterexamples but also even further away from measurement, for in addition to the components of well-being it now has to specify whether these components do or do not enhance well-being in any given context.

My complaint then is that as theories become ever more intricate and general, their relevance to the question of value aptness of science

---

4. For examples see Benjamin et al. (2014), Beshears et al. (2008), Lindeman and Verkasalo (2005).
5. See Sarch (2012) for ingenious proposals as to the formal structure of multicomponent theories such as the objective list views. The options are to treat each component as individually necessary, or else as additive, or else as required at a certain threshold, and so on. In each case the measurement will look dramatically different, but no objective list theorist in philosophy has even broached this issue.

diminishes. While the original philosophical proposals about well-being regularly inspire scientific projects, the subsequent versions with modifications do not, because their operationalisability is becoming harder and harder to achieve. This is not necessarily a problem—after all, true well-being may well be unmeasurable. But epistemic access and population-level comparisons is the conceit of the normal science of well-being. So any philosophical proposal that refuses to play the measurement game need not be taken seriously for these purposes. Yet the science of well-being rightly cares about conceptual validity rather than just operationalisability of its constructs, and one way or another philosophy must play a role in development and justification of these constructs. How can it?

2.3. THE VENDING MACHINE

Let us attend again to the analogy I invoked in the Introduction. In her many works on the nature of scientific theory Nancy Cartwright and colleagues distinguish between a *Vending Machine* and a *Toolbox* view.[6] On the Vending Machine view, a theory contains within itself the resources for the treatment of any concrete situation. We take such a situation—say, a fridge sitting atop a sloped ramp to the moving van—we feed in the inputs about the fridge, the ramp, the ground beneath, into the formalisms of Newtonian mechanics, and the theory 'spits out' a representation. This typically enables scientists to predict and perhaps also to explain the movement of the fridge. And it is supposed to be a *universal* vending machine too: the theory can do so for any situation, not just for the fridge but also for a banknote flying in a windy square full of people. Cartwright urges that this ideal ignores the realities of modeling, idealisations, and approximations in science. Once these realities are taken into account, we see that theoretical knowledge is far from sufficient for representing the world, and we should not be so optimistic as to think that physical theories apply universally.

6. Cartwright (1999, p. 185); Cartwright et al. (1995).

Instead, Cartwright and her coauthors propose that scientific theory is a toolbox—not in the instrumentalist sense according to which theories are mere tools for predicting, manipulating, and systematising observations but in a realist sense that theories contain some but not all of the tools necessary for building models that represent real situations. These tools are incomplete and do not always work. In her other work Cartwright argues that theories often make claims about *capacities*—stable forces that mix with other stable forces to produce observed phenomena. Capacity claims are abstract, and in order to be applied they need to be concretised using many locally appropriate approximations and idealisations that are not themselves theory motivated (Cartwright, 1989, 1999). This is the sense in which theories are toolboxes rather than vending machines. And this is the case not merely in practice or because we cannot do better. Cartwright believes that, for all we can tell, such a patchwork, or in her word 'dappledness', reflects the way the world really is.

For my purposes we do not even need to go that far. Cartwright's pluralist metaphysics are not necessary consequences of her views on methodology. But the 'spitting out' aspect of her vending machine analogy is instructive. The theory supposedly spits out representations of empirical phenomena. It does so using bridge principles that link theoretical concepts with measurable quantities. Adapted to well-being, the Vending Machine view encompasses the following commitments:

1. There is a single general theory of well-being.
2. This theory justifies the adoption of particular constructs and measures by implying them under certain assumptions about the context of research.

For example, say the correct theory is some version of idealised subjectivism—well-being is that state that a properly informed agent desires or believes to be good for their lesser informed self. Consider QUALEFFO, a questionnaire developed by the European Foundation for Osteoporosis to measure well-being of people with vertebral fractures and bone disease (Lips et al., 1997, 1999). It gauges their pain, physical, mental, and social function, and perception of health with 48 questions.

I have no view on how good of a measure QUALEFFO is, but suppose it does the job. According to Vending Machine, idealised subjectivism, plus assumptions about what sort of factors tend to cause (or correlate with) the valuable things specified by this theory in the population of people with vertebral fractures, implies that well-being of this population should be conceived approximately as the QUALEFFO indicates. To secure this implication, the Vending Machine theorist trusts that there should be a single theory of well-being and that this theory has a direct and straight-forward connection to the phenomenon measured by QUALEFFO.

I hope the story in the previous paragraph strikes the reader as far-fetched. Let us start with (2)—the claim that we can justify constructs by deriving them from one master theory. Whichever theory we pick of the ones currently on offer, this derivability looks utterly mythical. No one has the assumptions necessary to secure it. The bridge principles enabling the move from ideal desires to any item in the QUALEFFO are missing. Of course, we could make up these links: it is generally helpful to have healthy bones if one wants to pursue goals that one's ideal self would want one to pursue. Such a link can be made up for any of the existing theories of well-being, and they all sound perfectly plausible. But these links are not true bridge principles, for they are insufficient to *justify* QUALEFFO. Many things are helpful for pursuing ideal goals, but only some of them are included in this measure. Nothing remotely similar to this sort of derivation happens in real life and science, nor can we expect scientists to go through this process.

True, practical impossibility may not be a compelling consideration for philosophers who, for the most part, are happy to operate on the faith that the theories of well-being they so carefully craft could *in principle* be connected to constructs in the social and medical sciences. But I for one demand reasons to sustain this faith, especially when it so strains imagination. Indeed I see reasons working the other way. Current philo-sophical methodology worships different gods than those that would enable a connection between theories and measures. The philosophical gods are parsimony, universality, generality, immunity to counterexam-ples. When theories actually connect to measures in the sciences, these gods deserve no credit.

2.4. THEORIES OF WELL-BEING AS TOOLBOXES

The Toolbox theory, an alternative to the Vending Machine theory, does not maintain that standard philosophical theory is useless. On the contrary, it is essential, for it is a collection of models from which any act of measurement should start. Each model is a representation of well-being from a certain perspective, a conceptual tool to be deployed in a way it does not itself fully specify. Each of the Big Three regularly serves as inspiration for practicing scientists interested in well-being. Aristotelian views of well-being—that people have certain needs grounded in their nature—have inspired flourishing approaches in psychology and psychiatry as well as the capabilities approach in development economics. Classical hedonism serves as an explicit motivation for Kahneman's research program, among others. The subjectivist idea that realising one's priorities is important is often invoked in defense of life satisfaction approaches and of preference-based measures in economics. These are all examples of scientists feeding off philosophical ideas at various theoretical stages of their research.

The nature of this relationship is instructive. Scientists obviously do not treat the philosophical theories as providing the final and complete justification for their constructs. Their goals are opportunistic—they are just shopping for ideas, for inspiration. That there is a philosophical theory to go with their constructs and measures is a convenient fact, which, perhaps, provides an additional reason to take the constructs seriously. However, it is not a necessity, and there are other sources of inspiration and, crucially, as we shall see, other sources of justification.

Moreover, as far as scientists are concerned, picking up an idea from Aristotle or Bentham or Nussbaum does not commit them to the whole package. Psychiatrists and psychologists who study flourishing along dimensions of autonomy, mastery, purpose, connectedness, and other virtues dutifully acknowledge their debt to Aristotelian theories. But they also insist that the extent of flourishing be assessed by subjects themselves in questionnaires, thus injecting a hefty dose of subjectivism into their approach.[7] Life satisfaction advocates, similarly, mean *felt*

7. See Ryff (1989), Huppert (2009), Huppert and So (2013), Ryan and Deci (2001), Deci and Ryan (2008), and among others.

satisfaction where philosophers in the desire fulfillment tradition often insist on *actual* satisfaction of desires. And we have already seen that the best measures of hedonic state diverge greatly from the best philosophical formulations of hedonism.

The use of the capabilities framework in development economics is another case in point for the toolbox view. (See Appendix B for an explanation and references.) On its own, Amartya Sen's original idea that the notion of capability is preferable to utility is just an abstract framework that does not commit to much. Capabilities are distinct freedoms to pursue valuable achievements. Which achievements, whether they can be traded off against each other, and at what levels—all of these are substantive issues that need to be filled out before the proposal has any scientific relevance. Different projects fill it out in different ways. Compare, for instance, Nussbaum's (2000) robust Aristotelian version of the capabilities approach with its list of 10 valuable freedoms with the United Nations' much thinner Human Development Index that focuses only on life expectancy, income, and education (Anand & Sen, 1994). The capabilities approach is a valuable tool to represent a variety of outcomes—development, justice, freedom, sometimes well-being— not a vending machine.

The Vending Machine theorist bets that one theory of well-being underlies all measurement decisions. The fact that scientists themselves do not go through a derivation of these measures from some theory plus powerful assumptions does not show that such a theory does not exist nor that it is inert. The Toolbox theorist, on the other hand, sees the burden of proof on the Vending Machine advocate: unless you can show me these powerful assumptions, I will not believe in your vending machine. My sympathies are with the Toolbox theorist, though, I admit, an even-handed argument could go either way.

What is clear is that scientific practice conforms better to the Toolbox than to the Vending Machine picture. Unsurprisingly, scientists do not wait for a grand theory of well-being plus powerful assumptions before developing and justifying their constructs and measures. No research could get off the ground if scientists adhered to Vending Machine. However, the Toolbox view comes with its own costs. When a theory of well-being supplies only a few concepts and some inspiration, the existence of these tools does not uniquely compel any particular measure. And since the decision to use one tool rather than another appears to be a choice based on researchers' theoretical tastes, there is no justification of this choice either.

So we face the following dilemma: On the Vending Machine view theories justify constructs, but the view is impractical. The Toolbox view is practical, but on it theories do not appear to justify constructs. That would be bad news for a project on value-aptness of the science of well-being. Resolving this dilemma will require populating the toolbox with something more robust than just sources of inspiration. But before we get there, we should confront the possibility of a plurality of theories of well-being.

2.5. VARIANTISM VERSUS INVARIANTISM

Worries and scepticism about theorising are not new in philosophy. Having listed several 'fixed points' about what well-being is (that it must involve success in one's rational aims, that its experiential quality matters, etc.), Thomas Scanlon goes on to express doubts about the possibility of any more articulated theory of well-being:

> It does not seem likely, for example, that we will find a general theory telling us how much weight to assign to the different elements of well-being I have listed: how much to enjoyment, how much to success in one's aims, and so on. I doubt that these questions have answers at this level of abstraction. Plausible answers would depend on the particular goals that a person has and on the circumstances in which he or she was placed. (Scanlon, 1998, p. 118)

Scanlon is also sceptical about the usefulness of such a general theory, because this one notion cannot play the three roles it has been assigned: to help individual deliberation, to regulate third-person beneficence, and to be a consideration in political theory. It is likely that we shall have to develop separate notions for each of these purposes, he writes.

James Griffin similarly questions the very existence of a single notion of well-being (or of happiness and quality of life, for that matter):

there is no single notion of "quality of life" to cover all the instances of values that do indeed contribute to how prudentially good a life is. Each of us can, of course, commandeer the term "quality of life" and use it to cover whatever part of the whole domain of good-life-making features we should find most attractive. Perhaps this has indeed already happened in the course of the history of philosophy. Perhaps, therefore, some of us are not disagreeing with one another over the nature of a "happy" life but speaking of quite different things. (Griffin, 2007, p. 147)

Simon Keller connects this diversity to the failures of the traditional intuition-based method of theorising about well-being:

Around the notion of welfare lie several interrelated but distinct ways in which we can assess a person or her life. We can ask whether she is happy, whether she is satisfied, whether she is successful, whether she is flourishing, whether she enjoys a high level of well-being, whether she is doing well, whether she is doing well on her own terms, whether she is well off, whether she is living a good life, or whether she is getting the things that are good for her. These are all opaque concepts, but it is clear that they do not all amount to the same thing. It does not take much effort, however, to take our intuitions about any one of these notions, present them as intuitions about welfare, and reject a theory of welfare because it fails to meet them. But no single theory could capture all of these ways of assessing an individual or her life. So for any theory of welfare, there is a way to make it look silly. (Keller, 2009, p. 664)

Finally, among the social scientists, Des Gasper urges us to follow

Amartya Sen's principle that interpretations of inherently ambiguous ideas should illuminate, not attempt to eliminate, the

ambiguity. We should not claim that there is only one true version. We should ask, for any [well-being] or [quality of life] evaluation: who is doing what to/for/with whom, when, where, and
why? (Gasper, 2010, p. 353)

He then proposes three dimensions on which studies of well-being
can fruitfully vary: whether it is based on public or private values,
whether it is measured on a population or individual level, and
finally whether it focuses on life conditions or on subjective states.
He insists that faced with the diversity of conceptualisations of well-
being we should

> understand them as having different roles and different occasions
> of relevance. We need to reflect on and then focus in scope accord-
> ing to our judgements on: purposes, roles and standpoint, as well
> as on values, theoretical perspective, and the adequacy and feasi-
> bility of the required procedures and instruments. (Gasper, 2010,
> p. 359)[8]

So I am not the only one flirting with fragmentation. But is there a
coherent proposal in all this? Some of the sentiments just quoted can
be taken care of by Contextualism of Chapter 1. But the complaints go
further than that. There might be no single concept *and* no single theory to tell us when each concept is applicable and what states realise it.
Though these commentators may not accept my gloss, I think the view
they are calling into question is Well-being Invariantism (WBI). WBI
encompasses two claims:

> WBI1: The concept of well-being concerns the most general
> evaluation of the value of a state to a person and not anything else.

> WBI2: The full substantive theory of well-being will specify the
> unique set of conditions that apply in all and only cases of well-being.

8. See also McGillavray and Clark (2006) on the variety of concepts of well-being in development studies.

The first claim is what I have called the Circumscription thesis in Chapter 1. Denying Circumscription is one way to make sense of the ambiguity of well-being terms and expressions. The second claim—call it Uniqueness—trusts that a single theory specifies the extension of the concept wherever it applies. (Naturally, this theory does so at a very general level without implying that only a certain specific kind of life can be good for a person.) By itself Uniqueness is not averse to the ambiguity of well-being terms, so it is conceivable to deny Circumscription while at the same time claiming that a single theory specifies the extension (or extensions) of this family of concepts. Together, however, Circumscription and Uniqueness do imply that a single substantive theory of well-being should yield the conditions under which a person is doing well in general. Adding to this picture the Vending Machine view implies that this theory should also tell us the conditions under which a measure appropriately tracks well-being.

Contrast WBI with a view that rejects both the Circumscription and the Uniqueness assumptions. Well-being Variantism[9] (WBV) makes two commitments:

WBV1: The term 'well-being' (and its cognates) can invoke either general or contextual concepts of well-being depending on context.

WBV2: No single substantive theory specifies the realisers of every concept of well-being.

Let us call the first thesis Concept Diversity—the contextualist proposal in Chapter 1 was supposed to make sense of that. Now I concentrate on the second thesis—Theory Diversity. I phrased it as the denial of the existence of a single theory of well-being powerful enough to cover all the contextual notions of well-being. But I did not say how many theories of well-being that leaves us with or which theories those will be. Some philosophers, including those quoted earlier, already toy

9. I owe the term 'variantism' to Doris et al. (2008) who use it in the context of free will. Another similar sounding but in fact distinct idea is Guy Fletcher's (2009) 'variabilism'—a claim that no good contributes to well-being always and everywhere. This view is entirely compatible with and indeed friendly to well-being invariantism.

with the idea that temporal well-being is best understood in subject-ivist terms, while lifetime well-being requires a narrative shape to life that may go beyond how the person feels (Kagan, 1992; Velleman, 1991). I mean to go even further. More theories than those two are needed for all the diverse purposes of the science of well-being. How many exactly I cannot say. One possibility is that we shall need one theory per each context; another is that a single theory can take care of several contexts. This issue is partly empirical, partly pragmatic, partly normative: the number of theories that the science needs will depend on how many contexts a community identifies as requiring a standard of well-being and on how powerful/simple/useable we want these standards to be. I certainly will not be putting forward any answers from the armchair. This inexactness notwithstanding, Theory Diversity can still be elaborated.

It can be formulated as a stronger philosophical version and a weaker methodological one. The stronger version takes Theory Diversity and hence WBV to be a claim about the nature of well-being, squarely in the province of meta-ethics. It is an attack on the very idea of a unified theory of well-being. There is not one concept and there is not one thing the many concepts refer to. On the gentler version, we put aside the deep questions of meta-ethics and treat WBV as only a methodological thesis about how best to approach the sciences. It may be that there is a single correct theory of well-being, but for the purposes of construct justification we need not assume one. Instead we make use of several theories, and the arguments that justify the use of one theory in one context and another theory in another do not bottom out in a unified master theory. Rather, they are local arguments. Methodological WBV does not infer Theory Diversity from the fact that science uses more than one theory—that would be taking science uncritically. Rather it is a view that the apparent lack of a powerful master theory does not prevent judicious and normatively appropriate choice of construct.

In principle, the weaker version of WBV could even be compat-ible with some kind of invariantism about well-being. For example, we might think that any master theory, though it may be uniquely correct, is not well-suited for addressing the concerns arising in the sciences.

This theory may fail to be useful because, for instance, it is too abstract and too thin to yield the context-specific notions necessary for science and policy. Or perhaps the necessary bridge principles are missing. Whatever general theory of well-being is correct, justifying local constructs is a context-specific affair in which the general theory plays only an indirect role.

Which version of WBV do I advocate? The answer is the methodological one, at the very least, because I do not see the prospects for a master theory of adequate resolution. It sounds attractive but it is just not here. The invariantist's best argument is that such a theory is necessary if conflicts between different theories in the toolbox—for example, happiness versus achievement—are to be rationally resolvable. Value Aptness would be compromised if there was no fact of the matter about which construct of well-being is right, so I agree with this requirement. But I am pessimistic that the master theory is the best way to resolve such conflicts in scientific practice. In Chapter 4 we see that when values conflict, securing objectivity of science takes a political resolution, not an appeal to master theory. So I would neither bet my money on such a theory nor worry if it does not arrive.

2.6. IN FAVOUR OF THEORY DIVERSITY

A methodological argument in support of Theory Diversity could go as follows:

> *Premise 1*: The philosophical toolbox of the sciences of well-being includes many, not only one, of the current theories of well-being.
> *Premise 2*: Depending on context, different contents of the toolbox play a role in different constructs of well-being.
> *Premise 3*: Constructs of well-being, at least sometimes, specify the constituents, rather than mere causes or correlates, of well-being.
> *Premise 4*: Constructs of well-being in the sciences, at least sometimes, do a good job picking out well-being in a given context.

Conclusion: So different states, as specified by different theories, constitute well-being in different contexts.

I take the case for Premises 1 and 2 to be made already. We have seen how different are constructs across sciences—some are purely affect-based, others purely judgement-based; some invite general evaluations, others only evaluations of particular domains; some include objective indicators, some do not; and which indicators are included depends on whether we are studying children or developing countries or caretakers of the chronically ill. We have also seen that philosophical theories of well-being are used opportunistically as sources of inspiration and as repositories of ideas about what well-being *might* be, ideas to be used in part and in part rejected or modified as researchers see fit. Theories do not function as vending machines.

The next step is to justify Premise 3. This premise is crucial because it forestalls a natural objection on the part of the invariantist: Yes, indeed, constructs of well-being are diverse, but this has no bearing on philosophy because they pick out aspects, or causes, or correlates of well-being, not its constituents. There are a great many factors that bring about, prevent, or indicate well-being. In some contexts some of these factors are more powerful than in others, and some sciences are better suited to study some of these factors than others. This is where diversity of constructs comes from, argues the invariantist. It is a consequence of (a) the complexity of the causal web surrounding the phenomenon of well-being and (b) differences in the interests and methodologies of scientists.[10] Gerontologists focus on physical frailty not because relative health is constitutive of well-being of the elderly but rather because health problems have the greatest impact on the well-being of the elderly and they are what gerontologists should study. Economists focus on income because this is the one factor in the causal web of well-being they are well positioned to study, not because it realises well-being in some context—and so on and so forth. For each troublesome construct the invariantist will claim that it is a component, a cause, or a correlate, rather than a constituent, of well-being.

10. Rodogno (2014) mentions this line of argument.

Over the years I warmed up to this argument. It is a plausible story that can lead one to conclude that variantism and invariantism are both empirically adequate and the choice between them is a matter of taste. Still the variantist can respond. The first step is to agree with the invariantist on two accounts. Yes, well-being, like anything else, can be measured via indicators rather than directly. Sometimes constructs, and especially measures, are made up from strong correlates or causes of well-being rather than its constituents. The examples of health-based measures are plausible candidates for this (Groll, 2015). And, yes, different scientific projects have different purposes, constraints, and methodologies, which could explain some of the variation in constructs and measures. But these two facts are probably not enough to make sense of the entire range of diversity we are observing in well-being research.

Besides, the reply the invariantist is articulating easily lapses into Circumscription. The Circumscriptionist believes that 'well-being' and related terms refer all and only to the general all-things-considered evaluation of the goodness of a state for the agent. Whenever we undertake a more contextual evaluation anchored in a particular practical purpose from a particular point of view, we are talking about something other than well-being, for example, economic or health-related quality of life. So for the Circumscriptionist well-being research is at most about aspects of well-being. I grant this is a possibility. However, the supposed advantage of this position seems only to make room for invariantism. Its cost is to reinterpret research on well-being as being something other than what researchers think they are doing. Gerontologists may think they are researching the well-being of the elderly, but in fact they are only researching their quality of life, which is perhaps a component or a cause of well-being but not well-being proper. Why not? Because well-being is that general value for the sake of which we pursue quality of life, the invariantist insists. This sounds like foot stamping.

The other line of defense is to ask the invariantist for an outline of the theory that correctly specifies the referent of the narrow concept of well-being and yet does it substantively enough for the purposes of being a good vending machine. What in particular are we to do with two millennia worth of disagreements between the Big Three? The invariantist can claim that there is a resolution after all, or else that the disagreements do not matter. The first option is basically a promise that a unification of the Big Three will succeed. The second option requires isolating some core commitments of the Big Three and then claiming that these commitments are sufficient for all the very practical purposes of the science of well-being.[11] I can see attractions of each option, but they are not overwhelming attractions.

Finally we have Premise 4, which claims that sometimes constructs of well-being used in the sciences do succeed in capturing well-being properly. The variantist needs this premise, for without it the argument just attempts to draw a conclusion about the nature of value from the existing scientific practices—a fine instance of the naturalistic fallacy. What justifies Premise 4? No general argument can be made in its support. In part it is a bet that some of the many examples I discuss in this book are more or less successful at representing well-being. Empirically minded philosophers and philosophers of science frequently make such bets to ensure that their theoretical proposals do not veer too far from the practices that require an explanation in the first place. So Premise 4 is really a constitutive rule for a certain kind of project in philosophy. If the science of well-being was completely off track, then I would be writing a very different book. But Premise 4 is not purely a bet. In Part II I forestall an argument that well-being is inherently unmeasurable and discuss conditions for successful measurement.

So we have a version of variantism about well-being, at the very least a methodological one. In my view it is no less plausible than invariantism, though I admit the case is not exactly a slam dunk. Variantism may sound radical to a meta-ethicist, but for philosophers of science it should not be that big of a deal. They have realistic expectations about the power of theories as compared to more localised sources of knowledge such as models and mechanisms, and in my view so should philosophers of well-being. There is nothing exotic or radical about this

---

11. See Keller (2009), Tiberius (2007), and Bishop (2015) for attempts at unification and Taylor (2015) for prospects of a theory-neutral science of well-being.

perspective. There are several pluralist views in current philosophy of science. Concepts of species, innateness, genes, and others, it has been suggested, do not admit of a stable meaning, nor is a single substantive theory of these, both informative and true, possible.[12] Sometimes these philosophers conclude that the concepts are still useful; sometimes they advocate their elimination. Mark Wilson (2006) recently went as far as arguing that this ambiguity and fragmentation is indeed the normal behaviour of scientific concepts. If we take seriously the practical minutiae of their use and application, we see that even familiar notions such as colour, weight, and hardness only show stable content and orderly behaviour in patches, not universally, and only when they are anchored by reliable measurement and inference procedures (Wilson, 2006). An invariantist might respond with an incredulous stare and ask how, in the absence of a master theory, can the variantist know which tool to pull out from the toolbox? The variantist shrugs: in the same way as many concepts in science function. We start with a goal to study the well-being of a certain kind of creature, we know something about that kind, we reach into the toolbox to check if a theory of well-being for that kind exists already and if not we build it from the available tools adjusting it as we go along. Instead of debating the relative advantages of their views, our variantist and invariantist would do well to focus on this process. What exactly is in the toolbox such that it enables authoritative choice of well-being constructs?

2.7. ENRICHING THE TOOLBOX: MID-LEVEL THEORIES

The Big Three surely deserve a place in the toolbox. Many versions of these theories connect fairly straightforwardly to measures of happiness, life satisfaction, and flourishing. But, as we have seen with QUALEFFO, this is not always the case. Consider two more measures of well-being that do not easily reveal their philosophical ancestry.

---

12. See Ereshevsky (1998), Griffiths (2002), Griffiths and Stotz (2006), among others.

The first one is the Genuine Progress Indicator (GPI), inspired by the Canadian Index of Well-Being. It calculates the costs of economic growth and subtracts them from the standard gross domestic product. When social scientists speak of 'theoretical foundations' of GPI, they cite models and theories about the impact of economic growth on ecology, resource depletion, social costs, and so on. GPI reflects these costs and is thus is thought to gauge well-being better than GDP alone. It is sensitive to human needs and to the needs of communities extended across time, which must include contact with nature and a sustainable use of resources (Lawn, 2003). Implicit in these arguments are value judgements about what matters for decent community living. Satisfying human needs matters, and these needs include material comforts, belonging, connection with nature. Money matters presumably because it helps people satisfy their basic needs, but at a certain point more money is no longer necessary and might indeed take away from other needs.

The second example comes from gerontology. Many aging people care for spouses with chronic illnesses. This is frequently a time of great hardship in the life of the caregiver, who is at increased risk of depression and complications with their own health. Gerontologists define the well-being of caregivers to encompass subjective well-being and also freedom from what is called Caregiver Strain (Visser-Meily et al., 2005). Caregivers are strained if their sleep is disturbed by the care recipient, if they are lonely and pushed around, if their old life is gone and replaced with arguments, worry, fatigue, and so on. Underlying this measure is a judgement that, even if a caregiver is satisfied (as a loving and dutiful spouse may well be), the stress and pain of caring is still great and must be captured.

Why is it appropriate to conceptualise the well-being of caregivers in one way and the well-being of community in another? In each example above there appears to be a justification appealing to the nature of the kind in question (community, or a caregiver) and some normative principles not obviously grounded in the Big Three. The toolbox appears to be fuller than I implied earlier.

This is because in addition to *high* theories we also need *mid-level* theories. They are mid-level in that they stand between the high theories and the scientific constructs. (See Figure I.1.) Mid-level theories of well-being are theories of well-being in a particular context—the well-being of children, of the elderly, of the chronically ill and disabled, of people in stressful jobs, of institutionalised children, of an industrialised country, and so on. In Chapter 1 argued that along with the all-things-considered sense of well-being, there are also contextual senses, in which evaluation is relativised to circumstances. Mid-level theories are precisely substantive theories of well-being in contextual senses. To the extent that the Good Samaritan, the friend, and the social worker judge Masha's well-being, they do so by relying on an implicit characterisation of the creature that Masha is. For them she is a member of three kinds (a stranger on the street, a friend, and a client), and three accounts regulate their judgement—accounts of well-being for members of specific kinds in the circumstances their encounter takes place.

More generally I propose to think of mid-level theories as sensitive to three considerations. The first is the nature of the kind—who are the creatures whose well-being is in question? While high theories are about well-being of beings (typically human) in general, mid-level theories concern well-being of kinds defined more specifically: toddlers, adopted toddlers, adopted toddlers with Down's Syndrome, a country, caretakers of an ill spouse, and so on. The kind should be as specific as the circumstances require. Just as there is no one way of carving nature into kinds, there is no purpose-independent recipe for deciding which mid-level theories to pursue. Once a kind is specified and we know some facts about it, these facts serve as constraints on the mid-level theory for that kind. For example, knowing whether members of our kind are able to evaluate their situation, we may or may not decide to gauge their well-being with subjective indicators.

Second, the context is made up by the nature of the inquirer. Who am I to the being whose well-being I wish to know? In this respect context is sensitive to relationships and obligations that obtain between the subject and the benefactor or the knower: typically the closer the relationship between them, the richer and more demanding the appropriate notion of well-being is. What does it take for Masha who slipped and fell on an icy pavement to do well? If a stranger Good Samaritan is asking, then not a lot—just being able to get to her destination reasonably comfortably. The coparent of Masha's future child will demand more. When an aid charity measures aggregate quality of life with just a few indicators, instead of a more demanding measure, they are doing so in part because of their limited obligations to the people they study. But not only that.

Third, the circumstances of the inquiry matter. What is well-being such that I can do anything about it? For example, the practical feasibility for a benefactor of a given course of action in a given situation also affects the appropriateness of the standard of well-being this benefactor should set. Other things being equal, the less the benefactor can do for the subject of their beneficence, the lower is the right standard of well-being. What sort of outcome is and is not realistic to expect, given the existing resources and constraints, will fix the appropriate contrast class and thus the correct threshold of well-being.

National well-being and the well-being of a caregiver—my two previous examples—are two distinct nongeneral concepts; each requires a mid-level theory of their own. The theory of well-being of caregivers must be particularly sensitive to the dangers of loneliness, loss of connection with their mate, health complications, and, of course, what will be done with this knowledge. So far as I know, such a theory exists only implicitly in the measures currently used in clinical and social work settings. A theory of the well-being of a nation, or at least the beginnings of such a theory, does exist (Haybron & Tiberius, 2015). Its authors urge that this sort of well-being is primarily a matter of reflecting the fundamental priorities of the citizens and their success in realising these priorities. GPI is a good measure to the extent that it does so.

Mid-level theories are context-specific in their scope and narrow in their applicability; they are tolerant of exceptions and aimed first at guiding measurement or, outside science, an informal estimation. Having laid my own variantist bet on the table, it should nevertheless be clear that the notion of a mid-level theory is also easily palatable to an orthodox invariantist about well-being. No need to sign up for the entirety of my story. A high theory, both sides should agree, can play a role by constraining the space of possibilities in which to pursue mid-level theories without nevertheless determining them. Mid-level theories can be recast in terms of the Big Three (or the Big One should it arrive), but such recasting, interesting as it may be to philosophers, is not what justifies them. Instead further constraints will come from the nature of kinds and the context of theorising.

This is just what I show with an example in Chapter 3. A small gripe concludes this one. The world has plenty of high and not enough of explicit mid-level theories. To the extent that any exist, they are low on the ladder of philosophical prestige, as compared to fixing counterexamples to high theory. 'Sorry, I thought you were a philosopher' tells me a listener after a talk on this material. This is an unfortunate reaction. Building mid-level theories is an exercise in complex systematisation of values, facts, and other constraints—a perfect use of philosopher's time and an honourable use too. With its success stands or falls the value-aptness of science.

Chapter 3

How to Build a Theory
The Case of Child Well-Being

- I was just going to ask what he understands by the phrase [child] well-being?
- Frankly this is a ridiculous intervention. If you don't understand what we are trying to achieve about the well-being of children in this country you should have a serious look at yourself.

Debate in Scottish Parliament

According to Stewart Maxwell, the member of Scottish Parliament who bristles at his colleague's question, child well-being is so obvious that it is insulting to even ask to spell it out. Anyone looking at contemporary philosophy and the social sciences would be forgiven for concluding that academics too find such definitions unnecessary.

In political philosophy there is a mature literature on children's rights and the resulting obligations to them of parents and communities. One class of these rights, known as welfare rights, are supposed to protect children's basic needs such as shelter, nutrition, loving care, and so on. There is also a growing engagement with the nature and the value of childhood.[1] A notion of child well-being is immanent in all these discussions. However, merely specifying children's basic needs, or the

---

1. On rights and obligations see Feinberg (1980), Archard (2011), Liao (2006), Wieland (2011), among many others. On welfare rights see Brighouse (2003) and Brennan (2002). On the nature of childhood see Shapiro (1999), Brennan (2014), Gheaus (2014, 2015), Macleod (2010), Brighouse and Swift (2014).

goods of childhood, does not amount to a systematic theory of what is good for children and why.

In the social, psychological, and clinical sciences, child well-being is one of the most measured and researched outcomes with its own professional journals and societies.[2] Much is known about the determinants and risk factors of children's mental health, educational achievement, delinquency, prospective functioning as an adult. Nongovernmental organisations and dedicated charities regularly issue reports comparing child well-being across countries and time.[3] Typically researchers put together a set of these indicators and label this set 'child well-being'.[4] But what is missing is an explicit theory of that which the label refers to. Indeed social scientists often shy away from theorising for reasons that vary in quality.[5] One reason is a common aversion to philosophising and a mistaken perception that it perverts the goals of empirical research. A more compelling reason is the danger of state interference with parental rights in the name of child well-being. Indeed, Stewart Maxwell from our epigraph is the sponsor of a new law granting responsibility for overseeing well-being of a given child in Scotland to a person other than this child's parent—the so called Named Person scheme. His interlocutor's innocent-sounding request to define child well-being is in fact an objection designed to bring out the concept's vagueness so dangerous for policy and law. But even this reasonable concern is no reason against building a theory of child well-being. Indeed now that well-being is the currency of policy debates, the danger is precisely the absence of such a theory.

Finally, the high theories in philosophy rarely mention children specifically and almost never explain how the Big Three can be adapted to children. As such there is much relevant work in the right conceptual neighbourhood but nevertheless no proper theory of child well-being. Outlining such a theory is the first goal of this chapter.

---

2. *Child Indicators Research* is the journal of the International Society for Child Indicators.
3. The first and foremost is UNICEF's State of the World's Children yearly report (UNICEF, 1979–).
4. This is a fair description of Land et al. (2001), a typical study in the social indicators tradition.
5. See Seaberg (1990), McAuley and Rose (2011), Morrow and Mayali (2009), Axford (2009).

The second goal of this chapter is methodological—to show the process of building a mid-level theory of well-being in the sense articulated in Chapter 2. There are three kinds of raw material to work with. In keeping with the Toolbox view, we can start from the Big Three. Adapting these theories to the case of children and choosing the winner would be a top-down approach. Second, we can generalise from the existing measures of child well-being used in social indicators research. That would be a bottom-up approach. Third, we can feed off accounts of normal child functioning in developmental psychology and philosophy of childhood.

None of the three is enough on its own and each ends up being useful. The resulting theory identifies child well-being as having two individually necessary and jointly sufficient conditions. The first one is forward-looking: children do well to the extent that they develop capacities crucial for their successful future. The second condition is present-focused: children need to engage with the world in child-appropriate ways—attached, curious, exploring. A mid-level theory, recall, must be practical and usable in addition to being plausible. Since the uses of child well-being are multiple— research, education, welfare, parenting—I cannot anticipate them all and to this extent the theory is incomplete. Still it is worth articulating a structure which can be adjusted to circumstances.

3.1. BUILDING FROM THE BOTTOM UP

Social scientists who employ the notion of child well-being do so from several different perspectives: the social indicators tradition exemplified by the UNIECEF reports, the child welfare approach used in policy-related work, and developmental and educational psychology.[6] The absence of explicit theory in these endeavours notwithstanding,

---

6. UNICEF (2012), Bradshaw et al. (2007), Rigby et al. (2003), Federal Interagency Forum on Child and Family Statistics (2012). See Raghavan and Alexandrova (2015) for further references, the history of these measurement efforts, and differences between the approaches.

researchers do make assumptions about child well-being. These assumptions reveal valuable information for the mid-level theorist—they reveal the core beliefs about child well-being among the experts. If there is a normal science of child well-being, these are its pillars. I propose to use this indirect information to formulate *constraints* against which to judge any candidate theory. Of these constraints there are plausibly five.

The first one is that child well-being is measurable. One encounters a great deal of scepticism about the existence of a perfect or even the best measure of child well-being, but nevertheless child well-being is thought to be sufficiently nonmysterious so as to be epistemically accessible and allow the sort of comparisons that, as we shall see in Chapter 5, are characteristic of measurement.

The second is that child well-being is multidimensional. Most measures consist of several indicators and range from minimal (the child's health, material situation, safety) to richer (learning, relationships, etc.).[7] To avoid reifying these measures we should not conclude that child well-being has not one but many constituents. After all, it could be that well-being consists in nothing but a positive mental state, which is best measured by a multitude of indicators. Nevertheless, it is incumbent on any theory to provide a credible explanation of the multidimensionality of measures.

The third feature is the importance of objective indicators. This is in contrast to studies of adults where subjective indicators mostly rule the day. Some measures of child well-being are slowly beginning to incorporate one or more of subjective indicators (Ben-Arieh, 2006). However, they do so with crucial differences from the adult measures: (a) only older children are asked to share their views (b) only on some aspects of their lives (children are asked 'does it hurt when parents fight?' but not 'are you satisfied with life?' or 'do you like brushing teeth/getting vaccinated?') and (c) only in conjunction with objective indicators. This does not mean that how children feel is unimportant, but clearly there is

---

7. The advent of a richer model of 'best interest', one that goes beyond the physical and embraces the psychological needs of the child, can be dated to the classic publication on social work by Goldstein et al., in 1973.

an assumption that their subjective assessments of their lives, at least the assessments that research can access, are less available and authoritative than those of adults. Again not much follows from this fact alone about the nature of child well-being, but any theory should be able to explain this feature of measures.

The fourth characteristic is the developmental nature of measures. Childhood is a time of rapid change in mental and physical capacities, so measures are sensitive to the fact that different indicators are appropriate for different ages and stages (Ben-Arieh, 2010). Again, we should not infer that different states constitute child well-being depending on their age. The same well-being constitutive property can be realised by different states as the child ages. But nevertheless a mid-level theory of child well-being should explicitly incorporate development.

The final and related feature to extract from current scientific practice is a certain *dual* focus of child well-being—dual in that both future and present count. The focus on the future is reflected in the justification given to objective indicators: they are important because they predict crucial outcomes in adulthood, both positive and negative. As we survey in Raghavan and Alexandrova (2015), stable and secure attachment to an adult during the early years protects future mental health, enables exploration, as well as self-regulation, communication, and ability to form relationships.[8] Objective circumstances such as access to health, safety, and schooling all predict valuable adult outcomes, while poverty, violence, and disruption of attachment invariably predicts bad ones—mortality, morbidity, suicide, risky sexual behaviours, criminality, and so on. The empirical evidence is overwhelming and robust.[9]

This forward-looking aspect of good childhood is sometimes called 'well-becoming': childhood is the time to develop, to grow and

8. Warren et al. (1997), Grossman and Grossman (2005), Shonkoff and Phillips (2000).
9. The sources are too numerous to list but among them are Hillis et al. (2001), Dube et al. (2001), Anda et al. (2002), Dube et al. (2002), Dube et al. (2003), Chapman et al. (2004), Williamson et al. (2002), Dong et al. (2003), Dong et al. (2004), Farrington (1995), Brooks-Gunn et al. (1997), Shonkoff and Phillps (2000), Leventhal and Brooks-Gunn (2003), Wachs (1999), Bradley and Corwyn (2000, 2002), Starfield (1982, 1992), Rosenbaum (1992), Brooks-Gunn and Duncan (1997), Galobardes et al. (2006).

to invest in the future, that is, to become. But equally the consensus insists that there is more to childhood than well-becoming. The here and now matters too. A child with a miserable childhood is not well *qua child* even if they catch up in adulthood. This is the rationale behind the drive to include indicators of present well-being whether or not they predict future outcomes. Enjoyment and play may be of this type. Although they also predict future outcomes, this is not the only reason to include them in measures.[10]

So here we see an implicit theoretical judgement that child well-being is dual: a childhood is good for us in part because it prepares us for adulthood, but it also has to be good enough independently of adult outcomes. Shortly we shall see that there are parallels to this duality in the literature on philosophy of childhood, and we will formulate it more precisely.

To take stock, we have found five core assumptions about child well-being in the sciences: *measurability, multidimensionality, partial objectivity, stage relativity,* and *duality.* These will function as constraints on a theory of child well-being. The resulting theory must either meet these desiderata or provide a good explanation for why these are important features of child well-being measures. In proposing these constraints I do not mean to suggest that the measurement literature is infallible and that any theory of child well-being must model itself on the presuppositions of measurement. The current measurement instruments may be radically wrong, but they nevertheless appear to represent a certain consensus among the experts. We should be open to revising this consensus, but it is irrational to ignore it for purposes of initial theory construction. There is not a lot of *other* knowledge to fall back on.

3.2. BUILDING FROM THE TOP DOWN

These five constraints will prove useful for adjudicating between the Big Three. I turn to that shortly but first, I inspect another crucial source of information also from philosophy, but another branch of it: it turns out

10. Both the term 'well-becoming' and the insistence that it is not enough are in Ben-Arieh (2010) and Qvortrup (1999).

that our fifth constraint—duality—arises also in the literature on the nature and goods of childhood.

3.2.1. Making Sense of Duality

What is a child? The negative conception of childhood—childhood as incomplete or deficient adulthood—was present in Aristotle's and Hobbes's thought and was articulated recently by Tamar Shapiro. Shapiro (1999) develops a Kantian approach, according to which childhood is a 'predicament' out of which one emerges when one becomes a rational being whose views are worthy of consideration in the political realm. Many recoil from such a stark description, but it is hard to deny apparently basic facts: children are exceptionally vulnerable as compared to adults and they have a distinctive capacity to grow.[11] The forward-looking element or 'well-becoming' I mentioned earlier is likely feeding off this very idea: childhood is a stage on the road to somewhere further. It is hard to deny that this is part of the story.

This negative conception has recently met its more positive match. Samantha Brennan (2014, p. 21) agrees that emerging out of childhood is valuable for the child, but disagrees that childhood does not have goods independent of this emergence. She then proposes a list of goods that are intrinsic to childhood. They include trust and unstructured play—ideas echoed by several others.[12] Anca Gheaus (2015), taking cue from recent work in developmental psychology, adds to Brennan's list further goods that appear to be available mostly and especially to children: due to mental plasticity children have cognitive capacities for more intense and more open-minded exploration of their environment and more vivid sensations than adults. She concludes: 'Childhood may be a predicament in some senses, but in others it is a privilege: the privilege of superior abilities to learn and experiment' (Gheaus, 2015, p. 20). Social scientists who deny that well-becoming is all there is to child well-being plausibly do so on similar grounds.

---

11. See also Brighouse and Swift (2014), Brennan (2014), Gheaus (2015).
12. See also Matthews (2010), Brighouse and Swift (2014), Gheaus (2014), Macleod (2010).

Together the negative and the positive conception of childhood yield a more precise formulation of the duality constraint. The essence of duality is simple: although future outcomes can count toward the goodness of one's childhood, child well-being does not reduce to adult well-being. The constraint is best appreciated as a pair of claims. The first one, inspired by the negative conception is, Well-Becoming; the second one, inspired by the positive conception, is Nonreduction.

Well-Becoming: A childhood is good for the child only if it prepares them adequately for the next stage in life (however their abilities and circumstances allow).

Nonreduction: Having a childhood that produces high adult well-being is not sufficient for high child well-being.

According to Nonreduction a childhood could be valuable for the child (because it gives the child the special goods of play, love, etc.), even if, once its low value for adulthood is taken into account, it is bad overall for the individual that this child becomes. On Well-Becoming a childhood cannot be good for a child if it only contains the special goods of childhood without also growing up.

The tricky task is to specify the right relationship between the two claims. Accepting both means that child well-being will need to meet two distinct necessary conditions. This invites counterexamples, especially to the necessity of Well-Becoming. What about children who, for all we know, will not grow up to have any future at all, say, terminally ill children? Is it good for a dying teenager to learn to manage her finances?[13] Can't she have a good childhood just being a child?

One possibility is to stand ground: it is good for her to continue to develop in the remaining time no matter what. The other possibility is to relax the necessity of Well-Becoming for the special case of children who do not have a future to grow up into. This can be done by allowing trade-offs between Well-Becoming and Nonreduction. Children with no reason or ability to engage in activities that prepare for adulthood

13. I thank Caspar Hare and Richard Holton for this vivid example.

can still have well-being by sacrificing some forward-looking goods to present-looking goods.

Which of these options to adopt is not a question I intend to settle once and for all here. Gheaus (2015, p. 20), I submit, is right to argue that both types of goods—of childhood and of adulthood—are valuable for a person, and it could be rational to settle for several different combinations of the two. As a mid-level theorist I also have an additional worry. Allowing trade-offs in a scientific construct makes measurement all that much harder. Thus it would hurt my aspirations to formulate a theory that is genuinely usable. Still, philosophical considerations might speak in favour of trade-offs. Perhaps different *kinds* of children will need different versions of theories, which allow for different trade-offs. For now I formulate duality as a pair of necessary conditions, Well-Becoming and Nonreduction, bearing in mind that their joint necessity might need to be revised.

Now we are ready to apply the five constraints to the Big Three. As we test the main theories of well-being against the five constraints, we should keep in mind the distinction between high and mid-level theories. When a given high theory cannot provide resources for building a good mid-level theory given my desiderata, this does *not* speak against this high theory qua high theory. It merely shows that it does not provide the necessary raw material for this case. I argue that hedonism and desire-based theories are the wrong approaches to child well-being (which does not mean they would be wrong in other contexts). A version of objective list theory, on the other hand, is the way to go.

3.2.2. Hedonism about Child Well-Being?

What would a hedonist theory of child well-being look like? Hedonism, to remind, takes well-being to consist in the net enjoyment over the course of a lifetime. Childhood is one part of a lifetime, so a theory of child well-being is a temporal rather than a lifetime theory. Hence a hedonist theory of child well-being would be based on the net enjoyment *over the course of childhood*.

This formulation would satisfy Nonreduction. If a good childhood is an enjoyable childhood, as the hedonist says, this satisfies the constraint that a good childhood is good in part in virtue of its present quality, not just in virtue of its promise for an enjoyable adulthood. In keeping with Well-Becoming (the requirement that a childhood should prepare for adulthood), we could also add as a necessary condition 'a good promise for an optimal balance in the future'. So the duality constraint can be met by the hedonist.

What about the other four constraints? Mid-level hedonism can certainly satisfy measurability. As we have seen in Chapter 2, measurement of positive and negative mental states is in as good of a shape as ever. The extension of these methods to children, even infants, is hard but not impossible.[14] So let us grant the hedonist at least potential measurability.

Stage-relativity is also doable: enjoyable experiences that constitute good infancy may differ from those that constitute good toddlerhood, let alone good teenage years. Hedonism does not give a theoretical explanation of stage-relativity, but it can incorporate the developmental nature of child well-being by appealing to multiple realisability of enjoyable experiences—what is enjoyable to a toddler is not enjoyable to a tween, and so on.

Multidimensionality and partial objectivity, however, are far harder for a hedonist to accommodate. Why are measures of child well-being invariably multidimensional and invariably largely objective? A hedonist's response of this question will likely appeal to an empirical claim: children who fare well by objective indicators, that is, are safe, loved, cared for, stimulated in the right way, and so on, are the ones who enjoy their childhood the most. So multidimensionality and partial objectivity are merely features of measures, not features of the constituents of child well-being. But why should we take this claim on faith?

Institutionalised orphans provide a stark counterexample. Prospective parents who visit them and professionals who work with them often observe the equanimity, lack of crying, and blankness of their emotional state. Profound neglect and social deprivation no longer

---

14. See Casas (2011) on older children and Gopnik (2009, Chapter 4) on inferring mental states of babies.

bothers them because their bodies and emotional apparatus adjust to the fact that no one is coming to their rescue.[15]

The hedonist can reply that the childhoods of orphans are still less enjoyable than the childhoods of children who are not neglected. That may well be true, but it is equally true that children's emotional apparatus is not mature enough for their emotional state to properly reflect what is happening to them.[16] Because their emotional state is not fully informative on whether they are doing what children need to be doing (i.e., play, trust, learn, grow, etc.), we should not treat their mental state as the sole constituent of their well-being. This is why even when subjective indicators are used in measurement of child well-being, children are only ever asked very specific questions such as how much it hurts seeing their parents fight or not having friends at school.[17] So as a mid-level theory hedonism does not properly reflect what is perhaps the strongest intuition of parents, caretakers, and researchers: childhood is a critical time—the only time to actually develop, not just think we are developing, skills that are completely crucial for pretty much anything we will do later in life. So I move on to the next option.[18]

3.2.3. Subjectivism about Child Well-Being?

Subjectivism, to remind, is so called because it takes most seriously the agent's own values, desires, and goals. The realisation of those, whether actual or only some special superior set, constitutes well-being on this view. Can there be a subjectivist mid-level theory for children?

The simplest version of subjectivism defines child well-being as fulfillment of desires of this child. This version is hopeless from the start. It assumes an intellectual capacity to form, to order and evaluate goals, and to plan, tasks that require a high order of executive

---

15. Nelson et al, (2007).
16. Schore (1994).
17. Children's Society (2013).
18. Anthony Skelton (2014, in press) rejects experiential views of child well-being because in an experience machine a child would not be able to enjoy actual play and actual loving relationships. This line of argument is entirely consistent with the one I adopt here. I only prefer to arrive at the same result via considering the five constraints from science.

functioning that a child's developing prefrontal cortex simply cannot support.[19]

But the subjectivist has another trick up her sleeve. She could define child well-being in terms of desires of the adult that this child eventually would become. In particular, this adult's desires *about her childhood*. On this formulation child well-being consists in having the sort of childhood that an adult, perhaps a fully informed and a rational adult, would desire to have had. Given your own personality and priorities, what sort of childhood would you rationally want for yourself? That is the childhood that is good for you, the theory goes. Let us call this version *backward subjectivism* to mark the fact that it looks backward from adulthood to childhood.[20] How does it fare on the five constraints?

I suggest we grant duality to our backward subjectivist: the desires in question are about the childhood (thus taking care of Nonreduction), but presumably they could be informed by the adult's knowledge of what sort of childhood would have the best effect on their adulthood (thus taking care of Well-Becoming). But on the four other constraints, backward subjectivism fares very poorly indeed.

The epistemic accessibility and hence measurability of these good childhood constitutive desires is as poor as the epistemic accessibility of fully informed desires in general. Defenses of idealised versions of subjectivism bite this bullet. But a mid-level theory cannot afford to do so. Measuring priorities of adults about their adult lives is a live project, but to measure authoritative preferences about childhood as backward subjectivism requires is a taller order. Presumably the closest we come to it is when we ask experts on children—pediatricians, developmental psychologists, social workers, parents, and caretakers—what sorts of childhood it is rational to desire. My guess is that they would answer this question by just reciting the justifications of today's best measures and the findings of developmental research: children need to be safe, loved, cared for, fed, taught, stimulated, supported, allowed to explore, and so on. So it might seem that backward subjectivism passes muster.

19. See especially Welsh et al. (1991). Skelton (in press) similarly dispatches the subjectivist views because children's desires are not authoritative.
20. A suggestion by Matt Adler led me to formulate this option.

But it does not—not as a mid-level theory. A mid-level theory must not take empirical claims on faith. We simply do not know that all those good things experts pick out are sufficiently connected by causation or correlation with what fully informed adults desire about their childhood. There is no guarantee that a fully informed adult would not claim that actually for my sort of personality I would rather not have grown up in a family, thank you very much. This is the very same problem that critics of the full information theories of the good have originally raised.[21] A high theory can reject this criticism by appeal to superior theoretical virtues—subjectivism is the only theory that truly takes the agent's individuality seriously, hence we can pay some costs if the benefits are great enough. But a mid-level theory needs to have firmer empirical foundations. It has to be actually the case, for all we know, that rational adults would desire the sort of childhood that the best measures pick out. I do not know that and do not know how to find out.

This is also how backward subjectivism fails the constraints of multidimensionality and partial objectivity. It is perfectly conceivable that a fully informed adult would want their childhood to provide many, not one, goods, some objective and some not. But will this adult desire those goods that, given the best state of current knowledge, are good for children? Maybe, maybe not. As a mid-level theory backward subjectivism looks suspiciously high. In relying on empirical claims whose truth is unknown, it is unable *by itself* to evaluate the existing measures of child well-being. This is because this theory needs to use independent information, rather than information contained in itself, to tell us what to measure. Backward subjectivism can be made consistent with current measurement practices, but that is not good enough for a mid-level theory.

3.2.4. An Objective List Theory of Child Well-Being?

There are many objective list theories to choose from, but I cut straight to what seems to me the most promising version. According to Richard Kraut's *developmentalism*, well-being consists in flourishing appropriate

21. For example, Rosati (1995).

to the organism's nature and stage of development. '[A] flourishing human being is one who possesses, develops, and enjoys the exercise of cognitive, affective, sensory, and social powers (no less than physical powers)' (Kraut, 2007, p. 137). It is an objective theory with clear eudaimonist roots: well-being is the activity of a being to whom this activity is suited in virtue of their nature.

Developmentalism is a good starting point precisely because, unlike other Aristotelian approaches, it does not set a high bar on rationality. It is already conceived with beings other than adults in mind and does not require special moves to accommodate children. It is meant to apply to adults, children, animals, and plants, indeed any being that can flourish. Developmentalism passes our stage-relativity constraint without even trying—that goodness depends on stage in life is in its very definition. Since development is forward-looking, Well-Becoming also nicely accords with developmentalism.

What about the other constraints? Let us start with multidimensionality and partial objectivity. At first sight developmentalism looks to be in good shape. There are several powers a child should actualise as part of flourishing, and powers are by definition objective properties of a being. But that is far from where we need to end up. We need to specify *what* powers a child needs to actualise in order to flourish. Kraut enumerates cognitive, social, affective, and physical skills but purposefully does not say anything more specific and notes there is no mechanical procedure for making the list more precise. Importantly for Kraut, flourishing-constitutive capacities are not necessarily the natural or the evolved capacities. Some of these are bad to develop. For example, he claims that experiencing pain is bad for us (except instrumentally) even though it actualises the power of our organism to respond to harm. The same is true about our powers to inflict great harm on others. Instead, Kraut suggests, we start with some obvious examples (enjoying dinner with friends is good for us because it actualises our powers of eating and socialising) and watch them fall into a pattern:

> The argument is not that we have certain powers and inclinations when we are young, and therefore their development must be good for us. Rather, we notice as we systematize our thoughts

about what is good, that they fall into a pattern, and the notion of an inherent power waiting to be developed plays an organizing role in that process of systematization. (Kraut, 2007, p. 165)

The problem is that this method of generalising from obvious examples cannot take us as far as a mid-level theory requires. Which powers exactly are good for children? Other than appeals to common sense, Kraut's theory has no resources to answer these important questions.[22]

Another objection to Kraut's theory is that some things that are plausibly good for children are not primarily developmental goods. Skelton (in press) argues that such is love. It is good for a child to be loved not, or at least not only, because it develops this child's capacity for attaching and relating to people. Insisting that all goods of childhood are powers that this child develops seems artificial. This is an important objection since it shows how developmentalism struggles with Nonreduction. Nonreduction is a refusal to ground good childhood in the next stage of development. But development of powers seems inherently future-oriented.

I suspect there could be replies on behalf of developmentalism as a high theory, but I do not pursue them here. Instead I propose to reformulate the basic developmentalist proposal—that flourishing in childhood is in part growth—into a mid-level theory that meets the five constraints, without insisting that all prudential goods are powers to be developed and with a more robust criterion as to which powers count. This opportunistic approach is the essence of the toolbox methodology I described in Chapter 2. The theory described next readily acknowledges its developmentalist roots but also does not hesitate to twist and turn the original high theory given other sources of knowledge and constraints.

3.3. A THEORY OF CHILD WELL-BEING

According to a theory Ramesh Raghavan and I have already sketched (Raghavan & Alexandrova, 2015), children do well to the extent that they

22. Sobel (2011) mounts this critique in another context.

1. Develop those **stage-appropriate** capacities that would, for all we know, equip them for **successful future**, given their **environment**.
2. And engage with the world in **child-appropriate** ways, for instance, with curiosity and exploration, spontaneity, and emotional security.

But we have not defended it to my satisfaction. So here I pause on each of the conditions in turn, especially the central terms in bold, taking care to show how this theory meets the five constraints I articulated earlier.

3.3.1. Condition 1: The Future

Condition 1 is forward-looking in keeping with Well-Becoming. It specifies by reference to future outcomes the capacities a child needs to realise in order to have a good childhood. This, in my view, is the best way forward on the 'which powers?' objection to Kraut's developmentalism. There is, we have seen, a wealth of knowledge about determinants and risk factors of child and adult well-being at psychological, physiological, social, and environmental levels. Some of these determinants and risk factors sound like instrumental goods—safety and freedom from poverty. Those are plausibly properties of the child's environment, not the child themselves. Hence we may decide not to write them into the theory itself.[23] Others—being able to reason, make decisions, use one's body—are properly that child's capacities. They are noninstrumentally important for this child, because childhood is in part development oriented toward future outcomes.[24] They are properly part of the theory.

---

23. In general, however, mid-level theories are not beholden to keeping instrumental goods out of the core constituents of well-being. High theories normally dwell only with noninstrumental goods, but I see no reason why mid-level theories should, given their relative proximity to scientific practice.
24. Just because these powers are identified by reference to future outcomes does not make their development and acquisition instrumental. Development in childhood is not, or not only, of instrumental value.

'Stage-appropriateness' in Condition 1 makes room for the obvious fact that different children, depending on their age and ability, are honing different capacities. A child with a disability, even a serious one, can do that too. It is an empirical question which stage-appropriate capacities would be best in a given environment. At the population level, scientific findings on child development enable us to settle on a list of core capacities. In most environments of interest today, children need to learn to use their body appropriately, to communicate, to trust, to form and hold relationships, to learn about their environment, and, crucially to make decisions.[25] Normal development is highly stage-related; biological regulation and the ability to form secure attachment needs to be achieved in infancy; exploration, development of autonomy, and individuation are characteristics of toddlers; preschoolers learn to display initiative, self-reliance, and increasing amounts of self-control; children in primary school are mastering social norms and friendships; and finally, tweens and early teens are mastering higher order cognitive processes, emancipation, and self-identity (Sroufe & Rutter, 1984).

But Condition 1 need not be understood at the population level. It is formulated neutrally to allow for the importance of realising the child's individuality: it serves a child for their future to develop their unique endowments.

The next term in bold is 'successful future'. It refers to well-being either in adulthood or in whatever is the next stage in the child's development. Naturally it also reveals that the theory of child well-being depends on a notion of adult well-being or well-being in the next stage of life. This is unsurprising and should not be held against this theory. Children are kinds of human beings after all. Depending on the level of analysis—in particular on whether we are talking about an individual child or a large group of children—we might need a more or less detailed specification of the constituents of future well-being. We need not hold a theory of child well-being hostage to a full articulation of all

---

25. In discussing children's rights Brighouse and Swift (2014) specify four categories of forward-looking children's interests (physical, cognitive, emotional, and moral developments), a proposal congenial to mine.

other related theories, but it would be wrong to expect total independence between them.

Another crucial feature of Condition 1 is the relativisation of capacities to environment. This condition captures the uncontroversial fact that different cultural, historical, and social ecologies invite and enable different capacity realisation. Need a child know how to read in English and to play football? Presumably only if their environment makes that conducive to successful adulthood. Ethnographic evidence is rich with cases when formal schooling does not translate into adult well-being.[26] But this condition need not imply rampant relativism. Is it good for a little boy to learn to be an aggressive bully if that will give him a better chance in the future given his harsh environment? We can answer 'no' by specifying a robust normative conception of successful future. Aggression may not serve this boy well even if it enables him to become an unassailable overlord when he grows up. In this sense this theory may inherit the Aristotelian roots of Kraut's developmentalism in which flourishing is a deeply normative notion with both moral and prudential constraints. All theories, not just those for children, face the difficult question of whether or not moral goodness bears on well-being. So we should not expect a mid-level theory to come down on this issue once and for all—only if it arises in specific practical contexts.

Note finally the counterfactual and epistemic formulation of Condition 1: 'would, for all we know'. This serves to capture the uncertainty of what the future holds for children: even if they never live to have a successful future, it is good for them to develop the capacities that would serve them well were they to grow up. A person who unexpectedly dies at 21 could still have fared well as a child on Condition 1.

What to make of the cases when the child has no future to grow up into, discussed in Section 3.2.1? On the current formulation in which both conditions are necessary, such a child can have no well-being since she fails Condition 1. That sounds extreme. Can't some children with no future fare *better* than other children with no future?

---

26. For a review of anthropological literature on ecological determinants of child well-being see Stevenson and Worthman (2014).

The case of children who cannot develop any capacities at all may lead us to consider introducing trade-offs between Conditions 1 and 2, just as we considered trade-offs between Well-Becoming and Nonreduction. In extreme instances of a child with no future we might abandon Condition 1 altogether—just being a child might sometimes be enough. I note, however, that these are very specific cases indeed, since even heavily disabled and terminally ill children can develop *some* future-oriented capacities that seem to enhance their well-being.

3.3.2. Condition 2: The Now

The point of Condition 2 is to accommodate Nonreduction, making sure that the child also has nonderivative goods of childhood. There is more to childhood than Well-Becoming: overscheduling, helicopter parenting, and pushing children into developing adult-relevant skills at the expense of their childhood have all been named as tragedies of modern children even if (a big if) they lead to successful adulthood.

My notion of 'child-appropriate ways of relating to the world' is designed to capture those goods of childhood that make childhood good for the child whatever future brings. Importantly, child-appropriate ways must encompass both psychological states such as wonder, awe, carefree contentment—a specific child-like happiness—but also behaviours and relations: curiosity, spontaneity, exploration, and the sort of emotional security that comes from healthy attachment. This is how this account preserves a hedonist insight that well-being must come with a positive psychological state. But I hesitate to make 'pleasure' or 'enjoyment' a separate condition of child well-being as hedonists or hybrid theorists typically do.[27] 'Pleasure' is just not a category used by developmental psychologists who study children.

Recently these scholars have greatly furthered our knowledge of what it is like to be a child in a way that goes beyond the deficit conception of childhood: to be a child is to be a constant learner, far more intense, open-minded, and comprehensive than an adult, to explore the

27. For example, Skelton (in press).

world through play by building physical, biological, and psychological causal maps, to make up imaginary friends, to respond with wonder and utter excitement all the while feeling something like 'being in love in Paris for the first time after you've had three double espressos' (Gopnik, 2015).[28] Admittedly, that latter phenomenology applies to giddy infants more than to dour teenagers. Nevertheless, Condition 2 can be adapted to different stages of childhood: emotional security for teenagers amounts to an ability to break away from a parent to explore their own identity, while for an infant it may be nothing more than going to the other side of the room to pick up a toy before climbing back on the parent's lap. They are still both instances of child-like ways of relating to the world.

I happily help myself to this knowledge. However, I do not attempt to exhaustively specify child-appropriate ways. They cannot be subsumed under a purely biological definition of youth. There are child-like behaviours that are not worth protecting. Institutionalised children that rock themselves to sleep and practice other self-stimulating behaviours that are adaptations to their environments of deprivation are behaving in perfectly child-like ways. But these ways are tragic, and a child who engages in them is not doing well, though perhaps they are doing as well as possible. A war-time child who is not exploring and who is clinging to a parent for fear of loss is also practicing child-like ways that are tragic even if they are useful. A former child soldier who draws violent pictures of his former life might be healing, but he is not flourishing. [29]

So the notion of 'child-appropriate ways' will remain what Bernard Williams and others since have called a 'thick concept', in which the normative and the descriptive elements are intertwined. Child-appropriate ways are those ways practiced by the young that are worth protecting because they make for a good childhood. Specifying a full list of them might be impossible, but in keeping with our Nonreduction constraint this notion is unavoidable.

28. I rely here on Alison Gopnik's (2009) survey of research that includes work of her own, her colleagues' and other developmental psychologists'.
29. These adaptive and maladaptive behaviours might however qualify as well-being constitutive under condition 1 depending on environment.

Still we can say something more specific. Several of the existing proposals already mentioned speak to the valuable child-like ways of being. Brennan's (2014) list of 'intrinsic goods of childhood' includes unstructured, imaginative play, relationships with other children and with adults, opportunities to meaningfully contribute to household and community, time spent outdoors and in the natural world, physical affection, physical activity and sport, bodily pleasure, music and art, emotional well-being, physical well-being, and health. Brighouse and Swift's (2014) list includes, among other things, sexual innocence. Gheaus (2015) adds open exploration and exhilaration unburdened with previous knowledge.

The exact content of Condition 2 can be left incompletely specified at this stage of inquiry. Though I have aimed at a mid-level theory of child well-being, it is no doubt too high, that is, not applied enough, for some projects in the sciences of child well-being. Depending on the scale of the study (how many children are in question), the scope of the study (how many aspects of their lives are in question), and other factors no doubt, we will need further, more specific conceptualisations of child well-being. For lack of a better word we might need a *low-level* in addition to a mid-level theory.

For example a low-level theory of child well-being could operationalise Condition 2 in more ways than one. I anticipate two sources of variation: cultural/historical and purpose-based. Historians and anthropologists of childhood find meaningful differences in the conceptions of childhood throughout history and cultures.[30] Allowing some historical and cultural variations in Condition 2 need not, and certainly should not, lead to full blown relativism. Similarly depending on whether this mid-level theory is used in large-scale studies or in an individual encounter with a therapist or a social worker, the list can be longer or shorter.

It is straightforward to see that our theory satisfies duality, stage relativity, partial objectivity, and multidimensionality. What about measurability? Better measurement of outcomes, especially for children who are vulnerable, was my coauthor's primary motivation for developing this

30. Heywood (2010), Lancy (2014).

theory, and we discuss some measurement implications in Section 7 of Raghavan and Alexandrova (2015). Roughly our verdict on the existing measures of child well-being is that they often focus on ill-being rather than well-being, put too much faith in isolated indicators, and do not represent the goods of childhood of our Condition 2 nearly enough. Our own theory has no special obstacles to measurement by comparison to other theories discussed here. Indeed, many important aspects of well-becoming—health, physical growth, education—are well-represented in the existing measures. But a proper operationalisation must await another occasion.

3.4. OTHER MID-LEVEL THEORIES

A theory of child well-being is an urgent need that would not have been so urgent if philosophers were more open to science and scientists to philosophy. I stand by the theory Raghavan and I settled on for children, especially given the dearth of alternatives.[31] But more than that I am keen to advertise the need for more mid- and low-level theories— well-being with chronic illness or disability; well-being in displacement, flight, and migration; well-being in unstable employment; community well-being; well-being in the changing climate; urban well-being; and as many other categories as life requires. Building these theories takes a real change of habit for a philosopher. Note how factors that look instrumental to well-being from the point of view of certain high theories (exploring one's environment with curiosity or learning skills valuable for the future) nevertheless make it into a

31. There is only one explicitly articulated alternative to our theory that I know of— Anthony Skelton's (2014, in press). Skelton argues that child well-being consists in happiness plus several worthwhile goods particularly important for children. They are love, friendship, intellectual activity, and play. He raises compelling objections to both hedonism and Kraut's theory, concluding that our best bet is with a hybrid theory. In terms of what goods are noninstrumentally valuable for a child we probably do not disagree. Each of these goods should find their way into our Conditions 1 and 2 without trouble. My goal in formulating this account, however, has been to say more systematically what unifies all these goods—that they are goods of childhood specifically and that some of them are developmental.

mid-level theory. This specificity is important for a usable theory of well-being. Never mind that it goes against the normal practice in philosophy to formulate a theory only in terms of the noninstrumental goods. To quote a remark by my colleague and psychiatrist Felicia Huppert: 'It often does not matter where you draw the line between flourishing and its pre-conditions'.

The mixing of the instrumental and the noninstrumental goods in a single theory also means mixing facts and values, for it is in part the empirical facts about the kind in question that tell us what is good for this kind. This raises two questions. The first one is how to resolve inevitable conflicts—for example, does helicopter parenting or religious upbringing hurt children? The theory spelled out here will not answer these questions conclusively, for much depends on exactly how the two conditions are operationalised. The second question is whether such an entanglement of empirical science and moral philosophy compromises the objectivity of measurement and science in general. These are my tasks in Part II.

ACKNOWLEDGEMENTS

This chapter is inspired by collaboration with Ramesh Raghavan; some ideas are joint.

PART II

TOOLS FOR SCIENCE

If you go back 30 or 40 years, people said you couldn't measure depression. But eventually the measurement of depression
became uncontroversial. I think the same will happen with
happiness.

Richard Layard, The Guardian

The only science of smiling describes muscle movement—the rest
is bullshit.

4thpartypolitics, The Guardian (lightly edited)

This book's focus is what I have called the 'question of value-aptness'—
how the science of well-being can aptly represent well-being. We are
halfway toward answering this question.

So far I have argued that the meaning of well-being expressions is
sometimes unstable, that there is no master theory of well-being for
philosophers to hand over to scientists, and that instead of high theories we should pursue mid-level theories. These claims on their own
provide part of the answer to my main question—to be value-apt a scientific project should be clear about what well-being means in its context and it should be based on a theory of well-being properly suited to
the nature and the circumstances of the beings whose well-being is in
question.

But getting clear on the relevant concept and theory of well-being is not enough. Scientists also should be able to measure the property picked out by the right concept and described by the right theory. And they should do so while respecting a traditional albeit elusive ideal of objectivity. To show how this is possible, or sometimes impossible, is the task of Part II. My argument places me somewhere in between the two extremes in the epigraph, that is the optimism of Layard and the pessimism of the anonymous commentator on the Guardian website. I start by vindicating a possibility of objectivity even for a science as value-laden as ours and then vindicate a possibility of measurement, even valid measurement of well-being. However, the price of these vindications is a critique of a status quo, this time in science rather than philosophy, which pretends that objectivity and valid measurement can be achieved by deferring controversial value judgements to the 'people themselves'. This outsourcing secures neither objectivity nor validity of measures. To align science with values we need to do better. We need to put forward measures of well-being that are articulable and defensible to all involved.

Chapter 4

Can the Science of Well-Being Be Objective?

Consider a claim 'C causes E in conditions N' that is well confirmed by the standard methods of the scientific discipline in which this claim figures. What if the definition of C, or E, or N presupposed a moral standard such that this standard determined how C, or E, or N are conceptualised and measured? Should anyone trust this claim? Should they grant it objectivity? Should it be part of science at all?

So far I have argued that the science of well-being needs a certain kind of philosophy of well-being—sensitive to contextual variation in the notions of well-being and equipped with a number of mid-level theories of the states that realise these notions. This critique of orthodox philosophising aside, I am certainly not the only one to urge a joined-up practice of the philosophy and the science of well-being. Haybron (2008) advocates a new discipline—*prudential psychology*—his name for a union of philosophy, psychology, anthropology, and political science of human flourishing. Tiberius (2004, 2013) and Kristjánsson (2013) have also long advocated a better lining up between positive psychology and a duly rich theory of virtue and wisdom.

The problem, however, is that the typical conception of scientific objectivity is in tension with the efforts to link the science of well-being with its normative roots. Objectivity of science understood as value freedom in the content of scientific knowledge has been dominant in the twentieth-century philosophy of science. This conception is slowly losing its grip. But even as the layers of value freedom are being peeled off, there is still no positive story about how an inquiry such as the science of well-being could be both value-laden and objective.

Such a story is needed for many more projects than just the science of well-being. Empirical claims about health, child development, freedom, economic growth, resilience, frailty, and so on have the very same structure. They relate ordinary purely empirical variables such as geographic location with a variable that is defined in partly normative terms such as health status, as in 'Living in East St. Louis is a major health risk'. Or they may relate two variables that both appear to have a normative component, as in 'Unemployment negatively impacts well-being'. Health and unemployment as concepts are partly normative just as well-being is, in the sense that their definition and measurement, according to some anyhow, depend on moral claims such as 'Healthy life requires freedom from fear of being murdered', or 'Involuntary unemployment does not exist', or 'Happiness is necessary for well-being'. So in this chapter I broaden my focus from well-being only. I call the causal or correlational claims with such normative presuppositions *mixed claims* because they mix the moral and the empirical in a way that ordinary scientific claims do not. Mixed claims typically occur in the social and medical sciences such as economics and clinical and developmental psychology but can also be found in the biological and environmental sciences. Although philosophers have noted instances of such mixedness, there is still no clarity as to whether mixed claims should stay or go and whether they pose special problems for the objectivity of these sciences.

This chapter pursues four ideas:

1. Mixed claims are distinct from other well-rehearsed ways in which science can be value-laden (Sections 4.1 and 4.2).
2. Some claims in the social and medical sciences should remain mixed, against the advice to reformulate them into value-free claims or to move them outside science (Section 4.3).
3. The existing accounts of objectivity that make room for values do not fit mixed claims (Sections 4.4 and 4.5).
4. Nevertheless mixed claims can be objective in a sense that I articulate and defend (Sections 4.6 and 4.7).

I couch these theses as concerning mixed claims in general, but my examples are mostly about the science of well-being—not just because this is a book about this science but also because other normative concepts regularly bottom out in well-being. Measures of health, growth, development, and so on are justified in part on the basis of their ability to capture well-being. So an account of how the science of well-being can be objective will take us a long way toward understanding mixed claims in general.

4.1. WHAT ARE MIXED CLAIMS?

'Happiness is not always conducive to well-being' (Grueber et al., 2011).

'Long commutes are associated with lower well-being' (Diener et al., 2008).

'Early learning difficulties have a disproportionate impact on life well-being' (Beddington et al., 2008).

Social scientists who make such claims rely on a conception of well-being. This conception is reflected in their choices of a construct. (A construct, to remind, is an attribute, often unobservable, that serves as a dependent or independent variable in an empirical hypothesis.) As we have seen in psychology there are currently roughly three constructs of well-being. The first is a revival of a classical hedonist proposal to treat well-being as happiness or a favourable balance of positive over negative emotions. The second tradition takes well-being to consist in life satisfaction, an individual judgement about how one's life is doing overall. Finally, a third approach speaks of well-being as of flourishing or good functioning, an ensemble of strengths such as competence, relatedness, sense of achievement, and meaning. Each tradition has its own questionnaires or other means of eliciting self-reports of either emotions, or life satisfaction, or performance in various domains of life. When psychologists settle on a particular construct of well-being, that is when a heavy-duty substantive value judgement is made.

Other disciplines that study well-being—sociology, medical and clinical sciences, parts of economics—display a similar dynamic. Some hypotheses are on the face of it value free, but they rarely exhaust the full intent of researchers. Economists learn about happiness in order to have a more faithful account of economic growth; sociologists are interested in dignity and well-being at work; developmental psychologists focus on the processes and risk factors that greatly affect children's future functioning.

For all such claims I propose a definition:

A hypothesis is mixed if and only if

1. It is an empirical claim about a putative causal or statistical relation.
2. At least one of the variables in this claim is defined in a way that presupposes a moral, prudential, or political value judgement about the nature of this variable.

The first part of this definition specifies mixed claims as causal or correlational claims typical in the social and medical science. (They are normally probabilistic claims relating more or less general kinds.) Such claims play a crucial role in explanations and policy planning. However, they do not exhaust the science of well-being or any other science for that matter. Nor are causal claims somehow more fundamental or more important than theoretical claims, tacit nonpropositional knowledge, images, instruments, and so on. So we can equally well have mixed theories, mixed measures, and more generally mixed sciences. I concentrate on causal claims for convenience only, without meaning to exclude other vehicles of knowledge.

The more crucial feature of mixed claims is in the second part of the definition, that is, their reliance on a normative judgement. Such a reliance occurs in two ways. First, scientists might adopt a given measure because they believe it reflects well-being better than other measures— an explicit normative judgement. Second, scientists might follow a set procedure for measurement or data collection—for example, they might collect data on reported satisfaction with life—but this procedure is part of a broad methodological decision adopted by the founders of the research program of which they are a member. In this case, adoption of a measure betrays an implicit normative commitment to the validity of this research program. Either way the outcomes of the process are mixed claims, whether explicitly or implicitly.

What sort of values make for mixed claims? Philosophers distinguish between cognitive values, such as simplicity, explanatory power, coherence, generality, and so on, on the one hand and noncognitive values, such as moral, prudential, political, or aesthetic values, on the other.[1] It is the second kind that figures in mixed claims. For science of well-being (and also plausibly health and child development sciences), the most relevant values are prudential, that is, about well-being. In other cases, such as claims about involuntary employment, dignity at work, or political legitimacy, the values presupposed are ethical and political.

Without identifying them as mixed claims, philosophers have noted normative content in concepts of efficiency, rape, spousal abuse, unemployment, divorce, inflation, aggression, health and specific diseases, and, of course, well-being.[2] My notion of a mixed claim captures these examples. What has not been done is to settle whether mixed claims should be part of science and if so what rules they should obey.

This focus should be distinguished from the broader project of understanding the nature and significance of 'thick concepts'. Ever since Bernard Williams (1985) coined this expression philosophers have referred to 'well-being', 'courage', 'kindness', 'care', and so on as thick, differentiating them from 'good' and 'right', which are supposedly 'thin'. Although the precise definition of thickness is elusive, it is meant to signal a certain union between descriptive and evaluative content in a concept. For example, 'well-being' is thick to the extent that it is a good

---

1. Longino (1990), where noncognitive values are called 'contextual', and Lacey (2005), among others.
2. On efficiency see (Hausman & McPherson, 2006; Nagel, 1961), on rape (Dupré, 2007), on spousal abuse (Root, 2007), on unemployment (Hausman & McPherson, 2006), on divorce (Anderson, 2004), on inflation (Reiss, 2010), on aggression (Longino, 2013), on health and specific diseases (Hacking, 1995; Hawthorne, 2013; Kingma, 2014; Stegenga, 2015), on well-being (Tiberius, 2004).

thing to have, but also to fare well is to have a certain amount of health, not to be depressed, lonely, and so on.[3]

Thick concepts have long exercised meta-ethicists and philosophers of language. They have argued over whether thick concepts undermine the possibility of a moral theory, expose the limits of the fact/value distinction, and create problems for cognitivism, while others reply that in fact these concepts are compatible with many different stances in meta-ethics.[4]

My intention is to discuss mixed claims in science while inheriting as few of these foundational controversies as possible. Some philosophers take moral claims to express facts and others do not. Either remains an option for mixed claims. Those who take mixed claims literally will presumably treat thick concepts as referring to real entities with causal powers: for example, poverty, a phenomenon picked out by a thick concept, really does, on this view, have the power to cause heart disease. Those with more cautious meta-ethical views are free to adopt an antirealist reading of mixed claims instead: perhaps it is just a convenient manner of speaking to say that poverty causes heart disease. Either group should be interested in what I have to offer—ground rules for evaluating mixed claims in a scientific context.

4.2. MIXED CLAIMS ARE DIFFERENT

To formulate such rules I start by differentiating the value ladenness of mixed claims from other kinds of value ladenness. A taxonomy of the ways in which noncognitive values can enter science is interesting in itself, but its more immediate purpose is to show the uniqueness of mixed claims.[5]

3. See Kirchin (2013). Note also that my mixedness is a property of claims rather than concepts, but if we were to extend the property of thickness to propositions and not just concepts, then mixed claims would plausibly come out as thick. 'Someone who is well does not cry herself to sleep' would be an example of a thick proposition. I reserve the term 'mixed' for hypotheses and 'thick' for concepts.
4. See, respectively, Williams (1985), Putnam (2002), Blackburn (2013), and Roberts (2013).
5. This taxonomy is a product of conversations with Stephen John.

4.2.1. Values as Reasons to Pursue Science

To value knowledge, both theoretical and applied, is to value understanding and perhaps also the possibility of control over the environment. Without this normative stance the pursuit of science as a social enterprise makes little sense. But this sense of value ladenness clearly does not imply that individual scientific claims presuppose a specific standard about, in our case, well-being. It is at least conceivable to value knowledge without pursuing mixed claims.

4.2.2. Values as Agenda-Setters

Normative commitments about what phenomena are interesting, important, and worth studying are factors in setting the agenda for the sciences. For social sciences, Max Weber (1949) has famously accepted the role of cultural, moral, and other commitments for selection of one ideal type over another. Nowadays a similar argument is made by several others and not just about the social sciences. Hugh Lacey (1999, 2005) identifies *autonomy* as one of the senses of value freedom and defines it as the absence of external influence of moral, cultural, and economic nature on the priorities and direction of basic research. He maintains that such an autonomy is an impossible ideal, just because any scientific inquiry must start with a strategy that specifies what there is in the world to be known and how to proceed. Any such strategy starts from a cultural and historical standpoint and will prioritise some phenomena and methods over others by appeal to moral or cognitive values. A failure of autonomy, Lacey claims, need not necessarily compromise the authority of science. Philip Kitcher's (2011) ideal of a *well-ordered science* also calls for moral and political values, so long as they are endorsed by a community, to determine the agenda of scientific research.

Exactly how values should determine the agenda of science is a debate unto itself. For our purposes, we only need to distinguish this agenda-setting function of values from its role in mixed claims. There can be moral and political reasons to initiate a scientific study of human and animal well-being, but these reasons alone do not force mixing. Value-free definitions of well-being are perfectly conceivable, as we shall see shortly.

4.2.3. Values as Ethical Constraints on Research Protocols

A third and probably the least controversial role for values is the specification of ethical constraints on research. These constraints direct how to treat human and animal subjects during experiments, surveys, and clinical trials. Again nothing here speaks for or against the use of normative categories to define the target phenomena as in the case of well-being research. A scientific protocol can be ethical or unethical irrespective of whether the claims it produces presuppose noncognitive values.

4.2.4. Values as Arbiters between Underdetermined Theories

When empirical evidence alone is insufficient to adjudicate between two or more theories, values have been noticed, indeed called, to close the gap. Feminist philosophers in particular have invoked this argument to point out the legitimate role in theory choice of moral and political values (Longino, 1990; Kourany, 2003; and many others).

Our case is different. Take a mixed claim that long commutes are on average bad for well-being. This claim could, of course, be underdetermined by evidence. Is it really the commute? Maybe commuters are grim characters to start with? Confirming the badness of commuting for well-being requires a variety of intricate evidence: negative emotions, stress hormone levels, irritability, self-reports, and behaviour. Values, even noncognitive ones, could undoubtedly enter to adjudicate between equally confirmed mixed hypotheses. But crucially for us, this process is distinct from the mixed case: in mixed claims, say about well-being, the standard of well-being itself is not what closes the gap.

4.2.5. Values as Determinants of Standards of Confirmation

Another role for values explored originally in the 1950s by Richard Rudner and revived recently by Heather Douglas (2009) is in setting the level of evidence required for acceptance of empirical hypotheses. When there is uncertainty about a hypothesis that can inform important policy decisions (e.g., that drug X has certain side effects), moral considerations can be used to settle the level of evidence required for this hypothesis. Depending on the gravity of the consequences, a different level of evidence can be required. When the suspected side effect of the drug in question is as serious as a heart attack, even a small amount of evidence can be sufficient to accept the hypothesis that it causes heart attacks.

There is still a debate about whether or not such an importation of values into science is legitimate (John, 2015). But regardless of the outcome, the fate of mixed claims remains unaffected. Mixed claims can take inductive risks just as much as nonmixed claims. They would still remain value-laden even if moral considerations were purged from decisions about the required level of evidence.

4.2.6. Values as Sources of Wishful Thinking and Fraud

The history of science is in many ways a story of values entering into production of knowledge often in a way that serves the interests of the powerful. In our mixed cases, as we shall see in Section 4.4.1, these values too can determine what claims are accepted. But there is a *prima facie* distinction between clear wrongs such as fudging data, falsifying results, or rejecting a theory because it is Jewish on the one hand and basing science on thick concepts as in our case. It may still turn out that mixed claims are illegitimate, but that should be for a different reason than the illegitimacy of wishful thinking and fraud.

This completes our taxonomy for present purposes. There are other roles for values in science that may be confused with mixed claims. Values may also inform how we communicate scientific findings to the public or what metaphors we choose to describe them. But the bottom line is that mixed claims are in a class of their own—they need to be discussed separately.

In the Introduction I have already characterised the science of well-being as rejecting *neutrality*. To remind, according to neutrality scientific claims neither presuppose nor support social value judgements.[6] Mixed claims clearly violate neutrality. In the science of well-being in particular, mixed hypotheses presuppose a given standard of well-being and in doing so favour some conception of this value over another.[7]

Now we can ask the big question: Are mixed claims legitimate in science?

4.3. MIXED CLAIMS SHOULD STAY

The most explicit case against mixed claims can be found in Ernst Nagel's (1961) classic *The Structure of Science* in a section titled 'On the Value-Oriented Bias of Social Inquiry'. In it Nagel discusses the possibility that social science cannot, even in principle, be value-free. He cites Leo Strauss's examples of quintessential thick concepts—art, religion, cruelty—agreeing that the evaluative content is there and that it may be practically difficult to extricate it from the positive content. However, it is still possible if we distinguish between two senses of value judgement at play: one 'appraising' and the other 'estimating' (Nagel, 1961, pp. 492–493). We appraise when we endorse an ideal and judge something as meeting it or failing to meet it. We estimate when we judge to what extent a given phenomenon exhibits the features characteristic of a given ideal. Nagel's example is of anaemia, but let us apply his distinction to well-being. Scientists appraise when they take a stance on what well-being is and then use it to judge whether a person or a community is

---

6. Lacey (2005, pp. 25–26). Lacey (2013) eventually redefines neutrality as inclusiveness and evenhandedness, an ideal that mixed claims can satisfy, as we shall see shortly.
7. Jacob Stegenga formulates an instructive counterexample. Take a hypothesis 'Meditating about unicorns helps people fall asleep'. It involves a fictional concept, so call it a shmixed claim. Shmixed claims are empirical claims that involve both real and fictional concepts. Now, if we reason about shmixed claims as I do about mixed claims, then we would have to say that shmixed claims presuppose the reality of fictional entities. But that does not follow. The hypothesis can be a regular empirical finding while presupposing that unicorns do not in fact exist. In reply I reiterate that mixed claims do not presuppose reality of values; they are intentionally neutral on the question of metaethics. They do, however, presuppose an answer to a first-order normative question (say, whether well-being is happiness); this presupposition makes them mixed.

doing well. On the contrary, they estimate, when, using an account of well-being, they judge how much a person or a community exhibit the features this account deems well-being constitutive. In the first case, there is a genuine value judgement, while in the second a mere use of a normative criterion to make an empirical claim.

Nagel's goal in that chapter is a narrow one—only to establish that there is nothing inherently different about social sciences in the way they use values. For that, Nagel points out that physicists and biologists would also face the same issues when working with notions of 'efficiency' and 'anaemia'. I readily agree.

But his desire to prise apart appraisal from estimation has more ambitious roots. The point of drawing the distinction is to eliminate appraisal from science, leaving only estimation. The ideal science for him is an ethically neutral one (Nagel, 1961, p. 495). What I have called mixed claims are plausibly appraising claims on Nagel's picture. So his proposal, which is still endorsed by philosophers of science today, would be to reformulate them as estimation claims and eliminate the appraisal element.[8] How?

A natural way to implement Nagel's proposal is to convert mixed claims from regular causal or correlational claims into conditional claims. Take one of our early examples: psychologist Jane Gruber's claim that happiness is not always conducive to well-being (Gruber et al., 2011). Gruber documents the negative effects of positive emotions on problem-solving, social bonds, mental health, and so on. The title of her article—'A Dark Side of Happiness? How, When, and Why Happiness Is Not Always Good'—reads very much as an appraisal claim. But we can reformulate it as an estimation claim is as follows:

> If well-being is understood as good functioning across many domains and over the course of our lives, then happiness can impede well-being.

8. Heather Douglas (2011, p. 35) reflects on Nagel's proposal: 'The entanglements between the normative and the descriptive cast doubt on the possibility for any truly value-free statement of fact, but that need not mean we can have no objective statements. Central to such objectivity is the maintenance of at least a conceptual distinction between the descriptive and the normative'. Depending on how this maintenance is understood, her view may or may not allow mixed claims.

Since there is no commitment to the truth of the antecedent, this claim is value-free in the sense of Nagel's estimation claims. Nagel's position can then be summarised as follows:

> For any mixed claim involving a cause or a correlation C, a thick concept T and an operationalisation O of T
>
> 1. Scientists can investigate estimation claims: 'If T is operationalised as O, then C'.
> 2. Scientists cannot investigate appraisal claims that have not been so conditionalised.

I think Nagel's proposal should be rejected. While it eliminates values at one stage, it only pushes the decision about them to another arguably less appropriate stage.

Suppose we went with Nagel and reformulated mixed claims into estimation claims: there would still remain a question as to *which* normative standard scientists should use in their estimation claims. Which operationalisation should Gruber use in the antecedent? I can think of three answers a Nagelian could give.

The first one is to recommend that scientists stick to the proverbial folk theory of well-being. More generally mixed claims could be rendered value-free if they defined their thick constructs using the value judgements of the community they studied. 'Happiness can impede that which people call well-being', could be Gruber's claim. Or when both the putative cause and the putative effect are thick we get the following: 'What people call secure attachment is a major cause of what people call child well-being'. The problem with this proposal is that the folk disagree even within one community and any proposal for how such a disagreement can be resolved is itself normative.

The second Nagelian proposal is to counsel that scientists study the empirical relations between well-being and a given factor on *all* the existing views of well-being. If the folk intuitions are fairly represented by hedonic, life satisfaction, and flourishing approaches live in psychology, then the science of well-being should build up a store of conditional claims:

If well-being is positive hedonic profile, then it is caused by . . .

If well-being is life satisfaction, then it is caused by . . .

If well-being is a sense of flourishing, then it is caused by . . .

But it is hard to see why we should stop at these three. The history of philosophy, especially if we look beyond the Western tradition, boasts of other theories of well-being: well-being as knowledge of God, well-being as a meditative state, and so on. Using them all is impossible, but a choice requires a normative judgement about their relative plausibility—a judgement that the Nagelian hopes to keep out of science.

The third and most plausible Nagelian proposal is some sort of division of labour—philosophers take care of values while scientists take care of facts. The Nagelian would presumably argue that the right standard of well-being to use in the science of well-being is within the purview of moral philosophers (or perhaps the democratic decision-makers). Scientists can participate in this discussion but not *qua* scientists.

This proposal should also be rejected. First, in mixed cases normative decisions do *not* occur just at the beginning of the scientific process when the object of study is defined. Rather they keep reoccurring throughout, all the way down to the many practical decisions of scientific protocol. Those who define well-being in terms of authentic happiness need an account of authenticity and a whole string of other value-driven notions about how to measure it properly.[9] Some economists, as we have seen already, refer to the notion of 'clean' preferences to differentiate authoritative from unauthoritative desires. Definitions of child well-being refer to healthy and unhealthy parental involvement. When divorce is viewed as a transformation rather than only as a loss, it is worth studying the evolution of divorcees' coping strategies long after the divorce and not just their shock and loneliness immediately after (Anderson, 2004), and so on. On the separation picture, the scientist keeps running back and forth to philosophy (or keeps changing their identify from scientist to philosopher) whenever an evaluative question arises.

9. Sumner (1996), Tiberius (2013).

It is not the impracticality of this proposal that offends. After all, ethicists (or other specialists on thick concepts) could, on Nagel's proposal, be 'embedded' in a scientific process, for example as members of the lab who step in to make a normative judgement. Rather the problem with the proposed division of labour is that it ignores or devalues scientists' knowledge about values, which they have acquired in virtue of their knowledge of facts. This knowledge enables them to make better normative choices *qua* scientists. It is because developmental psychologists know the effect of, say, institutionalisation of orphans that they believe secure attachment to be crucial to child well-being. Similarly, it is because divorce scholars know the consequences of divorce that they conceptualise it as an opportunity for personal growth and not merely a loss (Anderson, 2004). In all of these cases value judgements are a result of an epistemic process; they are informed in part by facts and in part by the earlier value judgements made to detect those facts. Because of this process of co-evolution, scientists are in a good position to make some value judgements. Consulting philosophers and the public when making normative choices is important, but that does not mean scientists should refrain from using their own hard-earned normative knowledge.[10]

So the Nagelian division of labour ignores the methodological realities of mixed sciences and wastes the normative knowledge scientists acquire while in the business of producing mixed claims. This is a *prima facie* case that mixed claims are worth preserving.

4.4. THE DANGERS OF MIXED CLAIMS

What if mixed claims, defensible in theory, are dangerous in practice? They might bring with them dogmatism, bias, and wishful thinking. These are the very charges that have been levelled against proposals

---

10. To treat values as responsive to facts commits no meta-ethical sins. As Anderson (2004, p. 5) points out, even if Hume's prohibition of inferring facts from values is correct, values can still be supported by facts: 'Even if we grant that no substantive value judgement logically follows from any conjunction of factual statements, this merely puts

of feminist science and that advocates of feminist science have gone to lengths to deny.[11]

It is an empirical question to what extent mixed claims, as compared to nonmixed ones, foster scientific malpractice. I have not come across evidence on whether mixed claims are treated more or less dogmatically or whether its proponents are more or less likely to engage in wishful thinking. I readily allow for this possibility, but rather than speculating I concentrate straightaway on two well-documented dangers specific to well-being science.

The most serious charge is an importation into a science of substantive views about the nature of well-being that those whose well-being is being studied may have good reasons to reject. This danger is real. When eminent economists including Nobel Prize winners advocate a measure of national well-being that takes into account only the average ratio of positive to negative emotions of the populace (Kahneman et al., 2004b), the citizens can legitimately object if they take well-being to consist in more than that. Perhaps they believe that national well-being should also encompass the compassion, kindness, and mutual trust of their community, the sustainability of their lifestyle, not to mention justice.

A related danger is when the scientists engaged in mixed science fail to notice the value judgements they are making. Economists have been known for treating preference satisfaction and willingness to pay as definitional of well-being and thus not needing a justification. 'Cost-benefit analysis is what evaluation *means!*' said a UK Treasury official to a Whitehall civil servant (who reported this to me). In those cases presenting empirical findings about well-being, freedom, or health while failing to make explicit the normative assumptions on which these findings depend amounts to misusing the authority of science. It sneaks controversial values in through inattention.

value judgements on a logical par with scientific hypotheses. For it is equally true that there is no deductively valid inference from statements of evidence alone to theoretical statements. Theories always logically go beyond the evidence adduced in support of them. The question of neutrality is not whether factual judgements logically entail value judgements, but whether they can stand in evidentiary relations to them'.

11. See Pinnick et al. (2003) for critique and Anderson (2006) for reply.

Let us call these dangers 'imposition' and 'inattention', respectively. They are not the only dangers, but I submit they are the most visible and distinctive of well-being science. They undermine trust in it and raise the danger of coercive paternalism.[12] But instead of banning mixed claims from science, I propose to look for principles for their use that, though they may not guard against every danger, would at the very least guard against these two.

4.5. THE EXISTING ACCOUNTS OF OBJECTIVITY

A natural place to look for such principles is in the accounts of scientific objectivity friendly to values. As we shall see, they are of limited help.

Perhaps the best-known such account is Helen Longino's, summarised by herself thus:

> Data (measurements, observations, experimental results) acquire evidential relevance for hypotheses only in the context of background assumptions. These acquire stability and legitimacy through surviving criticism. Justificatory practices must therefore include not only the testing of hypotheses against data, but the subjection of background assumptions (and reasoning and data) to criticism from a variety of perspectives. (Longino, 2008, p. 80)

She argues that this sort of criticism can be secured by a community characterised by the following features: availability of venues for criticism, uptake of criticism, public standards to which theories and procedures can be held, and an equality of intellectual authority.[13] Like Longino, Hugh Lacey (2005) too emphasises pluralism of research strategies as a way of counterbalancing the value ladenness of background

---

12. For an argument to this effect see Haybron and Alexandrova (2013) where we define the notion of 'inattentive paternalism'.
13. Longino (1990, pp. 76–79, among other places).

assumptions. When scientific research proceeds from multiple ideological and metaphysical stances and when each is forced to justify itself in a public forum, the outcome is an objective inquiry, so the story goes. Douglas (2004) aptly calls this 'interactive objectivity'. I think interactive objectivity is not enough.

Pluralism and open criticism need a more robust formulation specific to the case of mixed claims. Otherwise these criteria are too vague for guarding against imposition and inattention.

Pluralism about definitions of well-being already characterises the science of well-being. As I have argued, no single definition of well-being dominates the current landscape. Such a variety of definitions could alert researchers to the problems of inattention and imposition. But by itself pluralism does not ensure that moral presuppositions are noticed and scrutinised in the right way. Measures of well-being are often selected for their ease of use, psychometric properties, institutional and disciplinary inertia, or personal preference. There is no guarantee in pluralism alone that these choices are noticed and called out for imposition and inattention. It is also not enough to say, as Longino and Lacey do, that different research programs need to be open to effective criticism. Mixed claims need a very specific sort of criticism on normative grounds, not just any criticism.

Another common criterion of objectivity—'impartiality'—faces a different problem: on one formulation it excludes mixed claims outright; on another it allows for mixed claims but without helping with inattention and imposition.[14] Impartiality specifies that cognitive values alone, and not moral and political ones, should justify our acceptance and rejection of theories (Lacey, 2005, pp. 23–24). To violate impartiality, it is claimed, is to commit the error of wishful thinking. Speaking of social sciences in particular Douglas argues:

> If values . . . serve as the reason in themselves for a theory choice,
> we have confused the normative and the descriptive in precisely
> the ways that Weber and Nagel warned us against. Our values

---

14. Impartiality is endorsed by Weber (1949), Douglas (2009), Lacey (2005), among others.

are not a good indication, in themselves, of the way the world is. (Douglas, 2011, pp. 23–24)[15]

The problem is when impartiality is formulated as forbidding that values determine acceptance of hypotheses, mixed claims face a test they could not possibly pass. This is because in a mixed claim the initial value judgement *does* preclude certain findings and to this extent values *do* determine what we will find.

Consider a stark example: a researcher is a staunch Aristotelian about well-being who believes that an immoral person could not possibly flourish, so she inserts a virtue constraint into her measure of well-being (ignore for a moment the practical difficulty of doing so). Using this measure, she finds that well-being is very low among sociopaths. Clearly this finding is determined in part by her initial value judgement and in this sense it fails the impartiality test. Similarly, psychologists who use life satisfaction questionnaires as their measure of well-being cannot discover a well-faring albeit constantly complaining kvetch, while those who use purely hedonic measures cannot discover a well-faring tortured artist no matter how satisfied she claims to be with her life. Definitions of well-being constrain the range of available findings, just as theories constrain the range of available observations. When this value judgement is part of the background theory, impartiality thus defined cannot be sustained. It makes mixed claims come out illegitimate by definition. They cannot escape the company of wishful thinking and scientific fraud. This is unsatisfactory: legitimacy of mixed claims should not be a matter of definition.

A better formulation of impartiality is as follows:

Impartiality$_2$: A claim is impartial if and only if, once all the value decisions about the constructs, measures, methods, and required

15. To be precise, Douglas's conception of impartiality is different from Lacey's. She does not rely on the distinction between cognitive and social values but instead on the distinction between direct and indirect roles of value. Once values have been invoked 'directly' in our choice of what to study and methodology, no further direct role of values is permitted. When it comes to confirmation of hypotheses values are only to be used 'indirectly' for managing uncertainty (Douglas, 2009, Chapter 5).

levels of confirmation are made, noncognitive values to do not play any further role in determining whether the claim is accepted.

To be fair, this is probably the version closest to what advocates of impartiality have in mind[16], and it may well be an acceptable version. Or, if it is not, I will not pursue this question any further. This is because impartiality$_2$ does not help us with the problems of imposition and inattention. This rule guards against the importation of values holding fixed all the other values. It cannot therefore tell us how to deal with those 'other' values so characteristic of mixed claims.

So while pluralism, open debate, and impartiality$_2$ are important, perhaps even necessary, ideals, mixed science needs additional principles.

4.6. OBJECTIVITY FOR MIXED CLAIMS

The additional principles we are looking for are partly principles of objectivity about values—prudential, ethical, political, whichever feature in mixed claims. To trust a science of well-being is in part to trust that it is based on an appropriate conception of well-being. But objectivity means (and has meant[17]) many things, so before I state the principles that realise it for mixed claims, I distinguish my focus from other objectivities.

Definitions of objectivity are not for the faint-hearted. Marianne Janack (2002) identifies no fewer than 20 senses of objectivity in contemporary philosophy of science alone. More manageably Douglas (2004) draws a three-way distinction: (a) objectivity as a way of 'getting at the objects' as they really are, (b) objectivity as a way of minimising bias, and (c) objectivity as a characteristic of the social process

---

16. Lacey (2003) makes allowance for the use of values in methodological choices in the human sciences, and Douglas would classify the choice of a thick concept to study as an initial methodological decision in which the direct use of values is permitted (personal conversation). Anderson (2004, p. 19) too is careful in formulating impartiality: 'If a hypothesis is to be tested, the research design must leave open a fair possibility that evidence will disconfirm it'(my italics), the implication being that choices of methodology are not always meant to be tested.

17. See Daston (1992) on historical shifts in conceptions of objectivity.

of science.[18] Each of the three senses mark a legitimate goal for the science of well-being. Focusing on (a), we might ask whether well-being is a plausible scientific object—is it stable enough to persist over peoples and histories enabling meaningful comparisons and theory building? Is it robust enough to changes in our instruments and methods? I began answering these questions in Chapters 1 and 2 and take them up again in Chapter 5. Focusing on (b), we might worry, as we already have in the discussion on impartiality, about dogmatism and wishful thinking.

While both of these foci are eminently legitimate, they do not help with imposition and inattention. Securing the right normative assumptions for mixed claims is neither a metaphysical task of making sure well-being is out there, nor is it a task of eliminating values. Rather I am after the sort of objectivity that ensures that values have undergone an appropriate social control, giving a community reasons to trust this knowledge. Such a control may not warrant blanket trust in a research project on the whole, but it would at least warrant trust in the project's value presuppositions and at least by the community that exercised control over these values. This sense of objectivity is closest to Douglas's (c) and to the 'procedural objectivity' that became popular in the twentieth century. Procedural objectivity focuses on the process of inquiry, not its results, aiming to ensure that this process is transparent, legitimate, and resistant to hijacking by specific individuals or groups.[19]

Historically, procedural objectivity has been thought to require value freedom understood as impersonality; that is, procedures should not presuppose the point of view of any particular person, group, or ideology. But value freedom and procedural objectivity do not stand or fall together. There could be good procedures for dealing with values. This is the conceit of philosophers who defend accounts of procedural moral objectivity inspired by the pragmatism of John Dewey. On this account the objective values are those that survive criticism in the public sphere and that are tested through 'experiments in living'.[20] Amanda

---

18. Douglas (2004) also draws further distinctions within each of the three modes, but they do not all concern us.
19. Porter (1995), Fine (1998).
20. Putnam (2002), Anderson (2014), Brown (2013).

Roth (2012, 2013) develops this proposal to show that an inquiry into values is procedurally objective to the extent that it is governed by principles that are really quite similar to the principles of science in the tradition of naturalised social epistemology. On this view value systems are treated just as scientific paradigms. They are confronted with empirical facts: 'You think teens should be taught only to abstain from sex? But what if that does not lower rates of teenage pregnancy?' Inconvenient facts can be explained away for some time until they can no longer be because that creates deeper moral problems elsewhere, and in this struggle the value system of a community changes.

This pragmatist story can be contrasted to constructivist conceptions of political objectivity that justify the principles of, say a liberal democratic state by appeal to outcomes of a more or less ideal deliberation.[21]Scientists who put forward and test mixed claims do not have access to ideal deliberation. This fact of life favours the pragmatist story for our case.

4.7. THREE RULES

But rather than entering the debate between pragmatists and ideal theorists, I move straight to those actionable principles that when used by a scientific community will deal with imposition and inattention. These principles are compelling whichever precise story about procedural objectivity is adopted.

4.7.1. Unearth the Value Presuppositions in Constructs and Measures

Inattention is a failure to acknowledge the values shaping a research agenda. Philosophers of science of all persuasions have urged that the first step to objectivity is making these presuppositions explicit.[22] I agree

---

21. Rawls (1993), Nussbaum (2001), Gaus (2011), among many others.
22. Weber (1949), Nagel (1961), Hausman and McPherson (2006), Douglas (2011), and others.

that this is the first step. Depending on the case, explicitness is more or less straightforwardly implemented. Sometimes all it takes is a sentence in the 'Methods' section of a journal article: 'In this study we assume that well-being consists in a favourable ratio of positive to negative emotions'. At other times, when scientific formalisms hide the value presuppositions, it takes a great deal of work, often a philosopher's eye, to uncover them.[23]

One reason is the sheer absence of an underlying theory in some cases. For example, measures of quality of life are often indexes constructed of several indicators. We have seen that in the social indicators tradition child well-being is measured by an index of infant mortality, vaccination, school attendance, and other factors. No researcher pretends that these factors *are* child well-being. They are only meant to be indicators of it. What then is child well-being? My efforts in Chapter 3 are motivated by the fact that this question is often left unanswered. But if explicitness is needed to combat inattention, and if inattention is an obstacle to procedural objectivity, such a failure to philosophise about the nature of well-being is a failure of procedural objectivity. So whenever scientists measure or otherwise study the well-being of X, they should be able to say, at least in outline, what the well-being of X is; otherwise they are not attending to their value presuppositions. In the next section I consider the possibility in which the precise definition does not matter because the empirical relation of interest holds on *any* of them. But even in this case scientists should be able to say what accounts of well-being the different measures presuppose.

As well as laying cards on the table, explicitness calls for an acknowledgment of alternative presuppositions, or at least awareness that they exist and that the disagreement about them is in part a substantive disagreement about values and not just a difference about which measures are more convenient. Though this aspect might take some scientists out of their comfort zone, explicitness is realistic to achieve. But merely having values in the open does not guard against imposition. The next two

---

23. Hausman and McPherson (2006, Chapter 2) is a classic illustration of how to unearth the moral assumptions in economic reasoning.

rules show what to do when disagreements about values arise, for example when relevant parties differ in their conception of well-being.

4.7.2. Check the Value Presuppositions for Controversy

Sometimes measures of well-being are 'robust' to fundamental philosophical disagreements. At their best, measures of child well-being, for example, attempt to capture conditions that if realised in childhood enable children to grow up happier, healthier, and more positively connected to others. Thus these measures stand up on all major theories of well-being that contemporary Western community entertains—experiential, subjective, and objective list. At least this is the hope. It is this robustness—an invariance to several different conceptions of well-being—that gives some mixed claims objectivity on the cheap, so to speak.[24] Unemployment has been repeatedly observed to lower happiness, life satisfaction, and mental and physical health significantly (McKee-Ryan et al., 2005). So a mixed claim 'unemployment lowers well-being' acquires mixed objectivity to the extent that it is true on several different conceptions of well-being.

But philosophising will not always be avoided so easily. Sometimes it matters a great deal which precise conception of well-being is selected. An example we have encountered already is national well-being. Different measures of it are not robust. A 2010 article by Kahneman and Deaton has a self-explanatory title 'High Income Improves Evaluation of Life But Not Emotional Well-Being'. Depending on whether scientists use life satisfaction measures (which capture evaluation) or happiness measures (which capture emotional well-being), they can reach substantively different verdicts on whether economic growth promotes

---

24. Not that cheap actually, because it requires a judgement of how strong the correlation between the various measures should be for the differences between these measures not to matter. Psychologists say that anything above $r = 0.4$ can be interpreted as evidence of either several or single underlying construct. Which interpretation is adopted depends on whether one is a 'splitter' or a 'lumper'. Nevertheless, Taylor (2015) gives reasons to expect some robustness of the choice of measure to fundamental theory.

well-being. Which measure is correct is a choice that in a democratic society should be made in a way that the next principle proposes.

4.7.3. Consult the Relevant Parties

When the choice of a measure of well-being is a choice between conflicting sets of values, consulting theory of well-being, high or mid-level, is unlikely to be the answer. I am sceptical that a plausible version of a theory of child well-being can say unequivocally whether tiger-mothering, or another kind of fundamentalism, hurts children. Where philosophy gives out, politics should step in. The only way to practice trustworthy science is to make this choice in a deliberative public setting in which the relevant parties are included. A measure of well-being that survives public scrutiny has procedural objectivity. Consider an example we have already come across.

Between 2010 and 2012 the UK Office for National Statistics conducted a country-wide inquiry called 'What matters to you?', soliciting views and recommendations from the public, the experts, and communities all across the United Kingdom (Office of National Statistics [ONS], 2012). Potential measures of well-being were released to the public and then respondents were asked to answer a few questions:

1. Do you think the proposed domains present a complete picture of well-being? If not, what would you do differently?
2. Do you think the scope of each of the proposed domains is correct? If not, please give details.
3. Is the balance between objective and subjective measures about right? Please give details.

The outcome of this exercise is a measure of UK's well-being that contains both subjective indicators—happiness, life satisfaction, sense of meaning—and also objective indicators, such as life expectancy, educational achievements, safety, and so on. A colourful wheel where each indicator is a spoke makes up for the fact that the measure basically includes everything but the kitchen sink. The ONS settled the seemingly intractable debates between the experts by including as many items into its final measure as practically possible and also by having the public vet this measure. No doubt the ONS measure has problems, but the honest effort to canvass the diverse views shows that the value presuppositions on this measure have arguably passed the sort of test I have in mind.

This example combines features of two relatively recent experiments in political science and science studies respectively: deliberative polling and systematic participation of the public in science. Deliberative polling occurs when a representative sample of the public comes together for a small group session with a moderator to discuss a question of public policy (e.g., should a minimum wage be required?).The participants receive input from the scholars who are experts on the topic via preliminary briefings. Moderators are trained to foster a respectful and inclusive debate. At the end the attitudes of the participants are measured and compared against their earlier attitudes. The precise conditions of good deliberation, its effectiveness, and obstacles are all topics of intense research in political psychology, the preliminary results of which are encouraging: participants respond to evidence, update their views, and become more politically involved and literate.[25]

While political scientists find ways of building consensus about politics, science studies scholars explore ways for people affected by a piece of science or medicine to have a systematic and nontrivial say in its methods, assumptions, or applications while at the same time respecting existing scientific expertise.[26]

Putting these two traditions together I propose 'deliberative polls of value presuppositions of mixed claims'. Groups of deliberators could be presented with various options for conceptualising well-being (or freedom, health, etc.) and with the relative advantages of each option normatively and practically. The deliberators will attempt to reach agreement according to whatever consensus-building and voting rules

---

25. James Fishkin, the head of the Center for Deliberative Democracy at Stanford University, describes the process in Fishkin (2009). See Myers and Mendelberg (2013) for an overview of empirical and normative research on political deliberation in small groups.
26. Chilvers (2008) and Douglas (2005) provide an overview of the history and the recent efforts.

they decide to put in place. Even if not everyone favours the values that survive such an exercise, the resulting consensus has some legitimacy and deserves trust at least from those whose views are admissible in a democracy and have been heard.

Such deliberations should include samples of all concerned parties. The ONS consultation happened by soliciting responses to an online questionnaire widely advertised through the ONS website, letters, and public events. Generalising from this example, I suggest that for mixed claims about well-being the deliberative polling should include

1. scholars of different approaches to well-being (plausibly philosophers, historians, or anthropologists).
2. researchers doing the measurement and data collection.
3. policy users of this knowledge.
4. a representative sample of the people who are likely to be affected when this knowledge is put into practice through policy, therapies, and other interventions.

The inclusion of experts is important because, as I argued in Section 4.1, scientists have knowledge of values in virtue of their empirical knowledge. Respecting this knowledge means that decisions about measures of well-being should not be outsourced to the nonscientists. But including the nonscientists is no less important. When scientists measure and monitor well-being this information can be used for oppression and domination. Science after all has the power and indeed a well-documented tendency to devalue nonexpert sources of knowledge.[27] Having people weigh in on how their own well-being is measured is a prudent reaction to these dangers, a reaction that need not assume that well-being is whatever people say it is.[28]

Can such exercises respect the expertise of scientists on well-being while at the same time avoiding imposing values on nonscientists? There

27. Wynne (1989) is a classic study of this phenomenon.
28. See Haybron and Tiberius (2015) for an argument that in a policy context well-being measures should be sensitive to the priorities of the citizens whose well-being is in question.

is definitely a tension. But much depends on implementation: what proportion of scientists to nonscientists is in the group, how the final consensus is determined, what checks there are on power imbalances. These are hard but not intractable questions. The deliberative exercises I sketch here are expensive, difficult to realise, and uncertain in their fruitfulness. So it is an open question whether mine is a rational bet all things considered. But it seems wrong not to try.

4.8. CONCLUSION

I proposed three rules: to make explicit the value presuppositions of mixed claims, to check whether the empirical claim is robust to disagreements about values, and finally, if it is not robust, to expose these values to an inclusive deliberation.

Together these three principles ensure that the science of well-being neither imposes values nor sneaks them past the people whose well-being is in question. Following these rules, I submit, secures procedural objectivity for the value presuppositions of this science. For other mixed claims, such as about freedom or health, these principles may need to be amended, but the spirit—objectivity as open vetting—should remain the same.

When the very definition and measurement of phenomena depends on moral categories, as they do in mixed sciences, we face a choice. We could reserve the notion of objectivity only for decisions and practices that avoid any such values. This is a view that preserves the neutrality of science at the expense of expelling mixed claims. I have argued against this. Mixed claims are already part and parcel of science. Pretending that they can be reformulated into value-free claims devalues perfectly good knowledge and stakes the authority of science on its separation from the community that enables it. The alternative I favour is to broaden our notion of objectivity to encompass also value-based decisions, such as which measures of well-being to adopt and when.

Chapter 5

Is Well-Being Measurable?

The last remaining element in my story is measurement. To be properly about well-being, it is not enough that the science adopts well-being constructs based on the right theories, nor is it enough that these constructs pass the objectivity test discussed in the last chapter. To be an object of science today, well-being also needs to be measurable and measured. In the Introduction I claimed that measurement is now a central pillar of the normal science of well-being. Gus O'Donnell's 2011 call to arms 'If you treasure it, measure it' was directed at policymakers and civil servants, but it was made possible by the comfort with measurability of well-being in the scientific community. Some of this comfort is very old indeed. Economics boast perhaps the most established tradition of measuring *welfare* as a function of wealth via various economic indicators such as gross domestic product, total surplus, household income, and consumption. But this is not the method that O'Donnell is celebrating and upon which the science of well-being stakes its novelty. Economic tradition of welfare measurement builds on the view that well-being consists in satisfaction of individual's preferences expressed in this individual's choices. My definition of the science of well-being is broad enough to encompass this approach. But the public face of the science of well-being typically comes with criticism of the economic definition and its replacement, or at least supplementation, with subjective well-being. 'Beyond Money: Towards an Economy of Well-Being' was the title of Diener and Seligman's seminal 2004 article. This is why more significant than the acceptance of the economic measurement of welfare is the growing acceptance of measures of happiness, life satisfaction, flourishing, and the many constructs in medicine that include the patient's perspective. They are now part of official statistics in many countries. As I mentioned in the Introduction, the fundamental disagreements about whether well-being can be measured have been replaced by specific disagreements about how and what aspect of it to measure.

But sceptics remain and should be heard. In this chapter I analyse what is to my mind the most compelling and explicit such challenge, both to the current measures of well-being, mainly the subjective measures that so excited O'Donnell, and indeed to the very idea that well-being is measurable. The challenge is roughly that well-being is too person-relative to measure reliably. I encounter this and related concerns often enough among the critics of the science of well-being.[1] But the formulation I attend to is due to Dan Hausman, and it appears in his recent book *Valuing Health* (Hausman, 2015). It is unique in that his focus is explicitly on measurability of well-being, rather than of happiness. As a psychological state, happiness's measurability is a matter of access to and comparability of mental states (one's own and that of others). So in debates on measurement of happiness philosophers focus on the ability of, say, questionnaires to gauge how people truly feel and to compare mental states across people.[2] Hausman, on the other hand, is concerned with well-being, which, as consensus has it, is a broader evaluative category, whose measurability consequently raises further problems in addition to the problems with measurement of happiness. In particular he argues that the concept of well-being calls for aggregation of goods in a person's life in a way that is duly sensitive to who this person is. The existing measures—whether focused on feelings of happiness, or life satisfaction, or quality of life—do not aggregate all the relevant goods in a way that respects individuality. So even if happiness is measurable, to the extent that other goods matter for well-being

---

1. McClimans and Brown (2012) and Hunt (1997) attack quality of life measures in medicine, respectively, for treating well-being as an outcome rather than a process and for not giving a clear definition of quality of life. Though distinct, these arguments perhaps echo Hausman's concerns.
2. Angner (2013) defends measurability of happiness as a psychological state. He grants that data from self-reports of happiness are easily misinterpreted as data about well-being but does not discuss whether well-being itself is measurable.

in addition to happiness, well-being will turn out to be unmeasurable. Comparing and ranking well-being states is possible albeit hard on an individual level, Hausman concedes, but becomes largely unrealistic on the population level required by science and policy. For him this means that well-being should not be the yardstick on which healthcare is evaluated and distributed and instead different categories, for example opportunities, are better suited for this task.

I start by reconstructing this argument and then argue that although it does not doom the project of measurement of well-being, it teaches a lesson for what sort of measurement should be expected and trusted. Given my variantist view—that well-being constructs can legitimately differ in substance and focus—one reply to Hausman is straightforward. Hausman selects the most demanding sense of well-being, one that calls for a comprehensive aggregation of all goods over the course of one individual's life. But this sense of well-being is not unique and may not be the right sense for science and policy. Showing that well-being is not measurable in one sense—I largely agree with Hausman on that—does not show that it is not measurable in all senses. Is well-being measurable in *any* sense relevant to science and policy? In my view Hausman's scepticism does not generalise and there is still hope for measurement of well-being in other senses. This hope depends, however, on abandoning the project of capturing the all-things-considered well-being of individuals and focusing instead on its commonly valued components or on well-being of kinds that share features and circumstances.

Second, I put pressure on a common critique of existing measures, which Hausman exemplifies. To undermine these measures critics tend to give intuitively plausible reasons why they should fail to capture well-being. But these intuitions can be very compelling and still fail to undermine a measurement tool if this tool systematically behaves in reliable ways consistent with empirical knowledge. The latter is the conceit behind *construct validation*, the main approach to evaluating measures in social and medical sciences, which Hausman does not discuss and which, I expect, most practitioners will appeal to in reply to him and other critics. So in addition to containing Hausman's scepticism, this chapter also presents an interpretation of construct validation. Construct validation, I argue, follows a coherentist spirit according to which measures are valid to the extent that they cohere with theoretical and empirical knowledge about the states being measured. In this sense construct validation is similar to validation of measures in other sciences, and to the extent that its theoretical ideal is realised in practice, measures of well-being can be valid.

So the overall message of this chapter is optimistic. Hausman's sceptical verdict is not warranted on the basis of reasons he gives and the science of well-being comes with plausible methods for checking that its measures are valid. Still my optimism depends on the science of well-being adopting realistic target concepts and implementing its ideal of measurement.

5.1. HAUSMAN AGAINST MEASURABILITY OF WELL-BEING

Hausman puts forward his case in the context of exploring whether health should be valued by its contribution to well-being—a question essential for deciding how a community should allocate its scarce resources for healthcare. How bad is it to have a broken ankle? A natural answer is that a broken ankle is as bad as the resulting loss of well-being to this ankle's owner. This is the view that Hausman rejects. One of his grounds is that a broken ankle can have a dramatically different impact on a person's well-being depending on who they are. In Hausman's own case a broken ankle enabled him to write the book making this argument. Such fine-grained heterogeneity cannot be accommodated by any realistic population-level healthcare policy. This, among other reasons, is why communities need to look for an allocation rule that is not based on well-being. As we shall see shortly, independently of healthcare concerns, heterogeneity is the main obstacle to well-being measurement more generally.

In order to make an argument against measurability of well-being, Hausman first needs to say at least in broad terms what well-being is. Although he does not articulate a full theory, the outlines are clear enough. Hausman believes that any account of well-being should accommodate the following constraints:

1. Well-being consists in several goods, not one.
2. 'What is good for me depends heavily on who I am' (Hausman, 2015, p. 121); that is, an agent's well-being depends on this agent's goals, values, and identity.
3. 'In assessing well-being we think primarily of whole lives, and our appraisal of how well someone's life is going during a limited periods often depends on what their life is like before or after' (p. 122).
4. Well-being is holistic in that adding more of some valuable good does not necessarily improve the whole. It's the combination that matters (p. 124).

Hausman is well aware that this conception is not entirely uncontroversial. Hedonists, for instance, argue that there is only one noninstrumental good—positive mental states. If well-being is directly measurable, these states are the only things that need measurement. Hausman does not hide his rejection of hedonism, and indeed subjectivism—neither bare feelings nor the fulfillment of desires or goals capture what it is to live well. In Chapter 11 he lays his cards on the table and backs a theory of well-being based on flourishing, aligning his views closely to developmentalism of Richard Kraut (2007):

> A fundamental evaluation of the value of some property or state of affairs for an individual depends on how the things that make human lives good (such as friendship, happiness, health, or a sense of purpose) are integrated into the dynamic structure of that individual's life. What Kraut and I call "flourishing" consists in the dynamic coherent integrations of objective goods into an identity. Well-being is flourishing. (Hausman, 2015, p. 141)

To argue that flourishing is not measurable, Hausman relies on a plausible conception of measurement—existence and epistemic access to a numerical scale that enables comparisons of all well-being states across and within persons and the distances between these states. This is known as an interval scale. Now it is easy to see how the case against measurability would go. Different people's flourishing is made up of different goods that combine in unique ways depending on their place in people's lives. No single scale applicable to all persons can capture the success of such unique combinations, so comparisons, let alone on an interval scale, are hopeless.

Hausman rehearses this argument but then recoils from its extreme scepticism. It is clear that sometimes comparisons are possible and very compelling—it is better not to die very young and not to become a child soldier, he agrees. It is equally true that sometimes feelings and preferences are decent guides to well-being. Adherents of idealised subjectivism, recall, hold that were an agent to form desires in light of full knowledge and with no mistakes in reasoning, these desires would be authoritative about what is good for this agent. Hausman is not a subjectivist, but he helps himself to the idea that some preferences are more authoritative than others. In particular, preferences can reliably indicate flourishing when these preferences are *laundered* in the right way. Laundered preferences are those held by individuals who are '(1) self-interested, (2) well-informed, (3) evaluatively competent, and (4) free of deliberative defects, and if (5) they have complete and transitive preferences among all alternatives' (Hausman, 2015, p. 132). When cases are sufficiently clear-cut or when preferences are sufficiently laundered, comparisons, even measurement, are possible.

This allowance notwithstanding, Hausman still ends up with a sceptical conclusion albeit less extreme. The reality in science and in public policy is that hard cases abound: who should get the scare resources—the ones with broken ankles or the mildly depressed? Plus the indirect measures of well-being—happiness-based or preference-based—are very poor at their task.

Hausman comes down especially hard against measures of subjective well-being. About life satisfaction Hausman (2015, p. 129) complains that reports summarising a great deal of information are unreliable and sensitive to irrelevant details. About measures of net affect, such as Kahneman's 'objective happiness', he notes that when the average affect at an instant is calculated all emotions get counted equally. He asks who decides how net affect is determined—why should my sadness at reading international news count for as much as my sadness at losing a grandparent? 'Heterogeneity goes all the way down to feelings', he insists (p. 129). Together these considerations show in his view that relying on subjective evaluation is too risky, because the precise impact of the quality of subjective experience for overall well-being is a personal matter.

Preferences, especially the laundered ones, would be in a better shape, *if* they were measured. But the fact is that standard economic methodology either infers preferences from choices people make (on the revealed preference approach) or else from their responses to questionnaires about what state of affairs they would prefer and at what rate (on the stated preference approach). Neither approach, he claims, makes an effort to select only the authoritative ones among these choices and judgements.

Here, then, is Hausman's tempered sceptical conclusion:

> Our evaluative abilities are limited with respect to our own lives, and the limits to those abilities imply limits to the completeness of our rankings of alternatives. It will often be the case that the objective of enhancing people's well-being does not discriminate among alternatives. As a practical matter, policy-makers will need other ways of comparing alternatives, and as a theoretical matter, either one has to conclude that prudence and ethics are less discriminating than previously thought or that normative notions other than well-being must play a large role. (2015, p. 142)

It is a tempered conclusion in that Hausman allows for uncontroversial comparisons of starkly different well-being states in individual cases—it is indeed better for him to lead the life he leads in Madison Wisconsin than to have become a child soldier. But in general these rankings will not be possible.

The following is a formalisation of his argument:

Premise 1: Well-being, being an inclusive good, allows for much heterogeneity in how and which component goods are integrated.

Premise 2: The existing measures are at best fallible indicators of some well-being relevant goods, but they do not respect heterogeneity.

Premise 3: Policy-relevant measures require a systematic population-level connection between well-being and the indicators.

Conclusion: Therefore well-being is not measurable for policy purposes.

The power of Hausman's argument is that, his endorsement of flourishing notwithstanding, it does not actually depend on this precise theory of well-being and can be accepted by proponents of different theories and even by those sceptical about a single theory. In Chapter 2 we came across Scanlon's (1998) worry that though we may manage to agree on core components of well-being (experiential quality, quality of life, success), a recipe for how much weight to assign to each is unlikely. This is Hausman's point too, but his additional contribution is to argue that this blocks social measurement.

Even some hedonists and subjectivists will agree. Premise 1 can be read as concerning instrumental goods, and no theorists of well-being denies that there are several such goods, nor that these goods can be good indicators of well-being. Premise 2 is also open to hedonists, for example, Roger Crisp, whom we encountered in Chapter 2. Recall that he takes well-being to consist only in enjoyment but rejects the possibility of objective measurement of enjoyment. If the x-axis represents time, the y-axis represents intensity, and a curve is formed from ratings of affect at an instant, then according to the hedonist tradition in psychology the total enjoyment is the area under the curve. Rejecting this picture, Crisp (2006) argues that an agent may well judge a given experience to be so high on enjoyment as to not be tradeable against longer experiences that are lower on enjoyment. This is a way of accommodating Millian high-quality pleasures and indeed Hausman's heterogeneity of agents. So his is an example of hedonism about well-being that is not committed to existing measures. Similarly, there could well be hedonists (or partial hedonists) who believe that shape of life matters in a way that makes it impossible to evaluate enjoyment at a time without considering the rest of the person's life. If enjoyability of an experience is time-dependent in a way that's sensitive to individuals' identities, a hedonist can again endorse the first two premises.

Similarly it is open to subjectivists to share Hausman's concerns about current measures expressed in Premise 2. Subjectivists, as we have seen, call for measures of well-being to gauge the extent to which agents' most important priorities formed under the right conditions are fulfilled. This is a tall order. Existing measures are unlikely to tap into laundered preferences. Merely asking people what they prefer and at what rate, or merely observing their actual choices in the marketplace, is a far cry from detecting the sort of authoritative judgements about deep values that subjectivists are after. I already mentioned that there are efforts under way to measure considered preferences through judgements and choices people make in carefully selected circumstances that plausibly reveal their genuine priorities—for example when medical students weigh the pros and cons of different residency programs and give reasons for their choices (Benjamin et al., 2014). I suspect these scholars would argue they *are* measuring what Hausman calls laundered preferences. Hausman does not discuss such attempts, but he might point out that health poses special challenges to preference-based measures—on what grounds would an agent make a laundered preference about the relative value of broken ankle versus mild depression? Evaluations of health states differ strikingly depending on whether they are offered by those who have experienced a condition or those who merely imagine it.[3] Like Hausman, subjectivists too may not be in a hurry to endorse these new measures: it is one thing to get people to form thoughtful preferences about residency programs, but whether these preferences are sufficiently close to the fully informed and the fully rational preferences that idealised subjectivists favour remains an open question.

All this is to say that Hausman's argument against measurement is far-reaching even in its tempered version. It does not require an endorsement of flourishing and would appeal to anyone who believes that the existing (and possibly any conceivable) methods of measurement of well-being do a poor job at detecting well-being in a way that respects

---

3. See Dolan (2000) on the relevant science and Carel (2016) on its significance.

differences between individuals. It is thus no good in reply to Hausman to just defend hedonism or subjectivism nor to attack flourishing. How then could we argue with Hausman?

5.2. IS WELL-BEING HETEROGENEOUS?

I start by questioning Premise 1—is well-being so heterogeneous as to doom measurement? Heterogeneity for Hausman means that there is no stable contribution a good makes toward all agents' well-being. Rather this contribution depends on the agent's history, values, and so on. It is the combination of goods that makes for well-being, not the presence of any specific good at any specific level. This appeal to uniqueness of individuals—no person is likely to have the same recipe for their well-being stew as any other—can be challenged. Are there really no commonalities to human flourishing? Tolstoy after all taught that all happy families are alike.

Consider happiness. In his *Pursuit of Unhappiness*, Dan Haybron (2008) bets that so long as this psychological state is understood in a sufficiently rich way, it is a good proxy for well-being. Happiness for him is no mere experience of pleasure, nor mere approval of one's life. Neither of these are deep or central enough to our psyche. Rather happiness is an emotional state that disposes us to certain ways of reacting to the world and to ourselves. Happiness for Haybron is the opposite of depression—the negative emotional state underlying all affects—not the opposite of superficial and peripheral affects such as pain or sadness. Three dimensions make up the happy emotional state: to feel at home with oneself and one's world (which Haybron calls attunement), to be engaged, and only lastly to endorse one's life. Think of the peaceful and awesome Santiago in Hemingway's *Old Man and Sea*. Call this Haybron-happiness.

Haybron contends that though well-being without this kind of happiness is conceivable, it is vastly unlikely—Wittgenstein famously claimed he had a great life, but who is inclined to wish such misery on a newborn? We are sure all unique, as Hausman says, but it is hard to imagine that we do not all need Haybron-happiness, at least to a minimal degree.

Depression is not compatible with well-being no matter what stew of goods uniquely suits each of us. Well-being must have something to do with 'living in accordance with our emotional nature' (Haybron, 2008, p. 193). Not all current measures of subjective well-being get at happiness so understood, but the emotional state in question is not inherently unmeasurable, and some measures come close enough (Haybron, 2016). Attunement, engagement, and endorsement are characteristics that, for the purposes of measurement, are similar to personality traits. If those are measurable—and it is hard to find more consensus in psychology than that traits of character are real and detectible—then so is Haybron-happiness. Should such a measure be available, it would plausibly tap into the most important and central aspect of well-being, so well-being would end up being measurable by a proxy.

Much heterogeneity in Hausman's sense will remain though, no matter how good our measures of happiness are. It is hard to function without Haybron-happiness, but how much health, relationships, fulfilling work, and so on we all need and in what proportion is likely to differ from person to person. Suppose there existed a measurement tool that allows individuals to judge (perhaps even authoritatively) the combination of goods uniquely suitable to them. The Organisation for Economic Co-Operation and Development (2013) Better Life Index has precisely such ambition when it invites people to rate various areas of life (work, housing, health, education, etc.) for their importance to them before evaluating their satisfaction with each. Whether such a tool will be practical and informative, especially for largescale science and policy, is another matter. Science and policy as practiced today require generalisations about components of well-being, and even if such generalisations exist, it is an open question whether they are sufficiently robust, scientifically and politically. They may hold the interest of scientists and indeed they do—hence the research on well-being at work, in cities, and so on. But policy robustness will be tricky—how important should a well-being component be for policymakers to be justified to invest in this good at the expense of others and with a danger of disadvantaging those individuals for whom this good plays a minor role? The comparability of the value of different bundles of goods will remain a problem.

For these reasons I am inclined to agree with Hausman that the judgement, let alone measurement, of the overall well-being of an individual is likely to be fiendishly complex—there are too many variables in the question of how a given unique person is faring in life, too many ways in which various goods may combine or fail to combine. This is why I do not pursue this reply much further and concentrate on a different challenge to heterogeneity—must the impossibility of measuring well-being in its most demanding sense doom its measurement in all cases?

5.3. CONTEXTUAL WELL-BEING AS MEASURABLE

The first thing to recognise is that while general well-being may be unmeasurable, some of its components are likely measurable. If well-being is a wholistic good as Hausman argues, it will be made up of some goods that are common to most or many individual packages. Haybron-happiness is one such component; others are health, positive relationships, security and so on. Multi-indicator measures of these surely provide valuable knowledge related to well-being, even for those who balk at calling them well-being'. But this is hardly controversial, and I believe more can be said.

Premise 1 embodies a demanding conception of well-being—requiring a complete aggregation of all important goods in a way that respects the agent's history, character, talents, culture, and values. In Chapter 1, I called this sense of well-being the all-things-considered evaluation and distinguished it from well-being in contextual sense. Hausman follows the trend in philosophy to call 'well-being' only the sort of evaluation that is relevant to a long-term personal therapist, a close friend, or an obituary writer. I have argued that sometimes 'well-being' connotes a less all-encompassing evaluation, such as when well-being is judged by family doctors, teachers, social workers, aid workers, and so on. In Chapter 2 I argued that the nature of this latter contextual well-being is the purview of mid-level theories, proposing one for children in Chapter 3. It is mid-level theories that underlie, though not as explicitly as I wish they did, the constructs in the sciences and the definitions of welfare in contemporary bureaucracies.

In these cases well-being is predicated of a particular *kind* of people in a specific type of circumstances. This sort of evaluation is at once narrower than Hausman's—not all goods are taken into account but only those shared by this group of people in these situations. It is also broader in that it considers a kind of person rather than an individual. A good social worker knows how to help families in crisis; a good child psychologist knows what troubled children need. A reader willing to grant this knowledge should also grant that it depends on modest generalisations about well-being, whether measured formally by indicators or questionnaires or eyeballed by an experienced specialist. There are examples in addition to those about child development in Chapter 3: recently adopted children benefit from a period of intense bonding with no one other than their parents; caregivers of chronically ill patients are at risk of ill-being even with social support.

Of course, if we focus on individuals we might find exceptions: recently adopted toddlers who can go to nursery school right away and caregivers who are just fine. But that is true about most generalisations, and the issue is ultimately empirical/pragmatic—how willing are we to attribute generalisable knowledge about well-being to specialists? It seems far more plausible to admit such knowledge about contextual well-being of kinds than about all-things-considered well-being of individuals. As I write this, thousands of unaccompanied refugee children from Syria, Eritrea, Afghanistan, and other troubled parts of the continent are roaming Europe—scared, vulnerable, abandoned. It is not a big mystery what is needed for *a* well-being of these children—security, stability, sustenance, and care. I say 'a well-being' because these goods are far more minimal than the child well-being I discussed in Chapter 3. Still such minimalism is defensible in an emergency, and in this sense the well-being of unaccompanied refugee children can be measured by their access to these basic goods. A table of indicators can represent the levels of each good, and, depending on how these indicators are aggregated, some comparisons between the levels of well-being of different groups of these children may be possible. Each of these children is an individual with a complex history, and it is possible to speak of the all-things-considered well-being of each. No doubt this is the sense their families worry about. As individuals, different children within this group will need different levels of security, stability, sustenance, and care, but a potential benefactor on a rescue mission is justified in ignoring these at least sometimes. Such a benefactor will speak about the well-being of these children as a group, and, because this group's well-being depends on a fairly obvious set of goods, measurement is conceivable.

If it makes sense to predicate well-being of kinds and not merely of individuals, then general claims about what is good for a given kind will be possible too. This is because kinds are identified by the generalisations they support—that is one common definition of kinds anyway (Boyd, 1991)—there will thus be generalisations about how members of this kind function in such and such circumstances.

To the extent that such knowledge is possible and to the extent that this knowledge is about well-being *in a sense*, we have another reply to Hausman. He selected the most demanding and the least epistemically accessible notion of well-being and showed an impossibility of measurement for this notion based on the intuitive impossibility of making generalisations about well-being. But this is too easy. Uniqueness of earthquakes, avalanches and wars does not stop their scientific study, and contextual well-being is no different. Hausman could retort that contextual well-being is not true well-being. It is perhaps quality of life, or performance according to one indicator, but not well-being proper. But at this point the argument has shifted into an unhelpful territory about who is entitled to the term 'well-being'. Erring on the side of liberality as I did in Chapter 1, I maintain that there is more to evaluation than judging individual lives all things considered.

Still, even allowing for well-being in this contextual sense, what confidence should we have in the existing methods of its measurement? It is not enough to show that Hausman's argument is premised on too demanding of a notion. That merely shows that there are other notions that apply to kinds, and, since kinds are based on generalisations, these contextual notions are better candidates for measurement. That secures potential measurability of contextual well-being. But to address Premise 2 we also need to show that measurement of contextual well-being is in actual fact realistic and defensible, at least more so than Hausman maintains.

5.4. CONSTRUCT VALIDATION

So far I have said very little about what measurement and validity are. To make further progress on Hausman's challenge, we need to attend to why scientists take the existing measures to be valid. I argue in this section that against these measures Hausman offers insufficient evidence—a mixture of appeals to intuition and unsystematic references to studies that expose problems in one or another questionnaire. But the field of social and medical measurement has elaborate procedures for validation. Evaluating current measures of well-being takes evaluating these procedures, not appeals to intuition.

If we asked scientists why they use a given measure of well-being, their answer would invoke *psychometric validation*. The psychometric tradition in the social sciences has traditionally specialised in developing tests and questionnaires for detecting intelligence, personality, and lately well-being. Some of the measurement tools and their use in research on race, gender, and class, especially in the early twentieth century, have an unsavoury history. The eugenic roots of this work are dutifully and solemnly acknowledged in the introductory courses to psychometrics.[4] But for virtually all researchers who measure an attribute on the basis of people's reports or performances in tests, psychometric validation remains the obligatory procedure. Large swaths of the science of well-being in particular have embraced questionnaires and with that psychometric validation.

5.4.1. Textbook Procedure

Validation follows a typical pattern described in measurement textbooks and articles.[5] First, researchers define the construct to be measured by elaborating its scope and limits. This is the conceptual stage in which meaning of terms is discussed, invoking anything from classical sources ('Aristotle said . . .') to untutored intuitions, to dictionary definitions. For

---

4. For a brief history of the first psychometrics laboratory see http://www.psychometrics. cam.ac.uk/about-us/our-history/first-psychometric-laboratory.
5. For example, DeVet et al. (2011), Simms (2008).

example, the scope of happiness is often deemed to be the overall positive affect, while the scope of satisfaction with life is a cognitive judgement about one's conditions and goals. In the second stage, researchers choose a measurement method (a questionnaire, a test, or a task), select the items (what questions? what tasks?) and settle on the scoring method.

The Satisfaction with Life Scale (SWLS), which we have already encountered and which Hausman specifically mentions, is a popular five-item Likert scale for measuring the cognitive aspect of subjective well-being, that is, the extent to which subjects judge their life to be satisfactory. The process of its validation is described in the much-cited article by Ed Diener and his colleagues (Diener et al., 1985). The first two stages in this case consisted in analysing 48 items all in the conceptual neighbourhood of subjective well-being. The team eliminated questions about affect because they viewed life satisfaction as a cognitive judgement about one's life as a whole. This left them with 10 questions, a further five of which were eliminated because of 'semantic similarity'.

In the third and final stage, the instrument is tested for its construct validity, that is, its ability to capture the intended attribute. [6] In the case of well-being measures, this stage frequently involves *factor analysis*: when hundreds of subjects fill out the same questionnaire, it is possible to observe the correlations between responses to different items. Unless the questionnaire in question is single-item, a rare occurrence in this field, much can be learned from these correlations. In particular they are used to show that there are one or more clusters of items called 'factors' that account for the total information. Scientists speak of factor analysis as extracting 'a manageable number of latent dimensions that explain the covariation among a larger set of manifest variables' (Simms, 2008, p. 421).[7] Here 'explanation'

---

6. I concentrate on construct validity at the expense of other validities because among measurement theorists the consensus seems to be that construct validity encompasses all other types of validity, such as criterion, predictive, discriminant, and content validity (Strauss & Smith, 2009). See Chapter 6 for a discussion of content validity.

7. There is a difference between *exploratory* and *confirmatory* factor analysis (see De Vet et al., 2011, pp. 169–172, among others). The former is used to reduce the number of items in a questionnaire by identifying the one(s) that best predict the overall ratings. The latter, on the other hand, tests that the factors that best summarise the data also conform with a theory of the underlying phenomenon if there is one. This distinction is not important for the present argument.

is evidently used in an entirely phenomenological sense as saving the phenomena (the phenomena being the total data generated by administering the questionnaire in question), rather than stating the causes or presenting a theory of the phenomena. For the SWLS, factor analysis identified all five items to be measuring the same latent variable because a single factor accounted for 66% of the variance in the data (Diener et al., 1985). Other scales may turn out to gauge more than one dimension.

The next step of the testing stage is to check that the behaviour of these factors accords with other things scientists know about the object in question. In case of subjective well-being, this knowledge includes how people evaluate their lives and surroundings, what behaviour results from these evaluation, and what other people who know the subjects say about them. For example, the SWLS, according to its authors, earned construct validity when Diener and his colleagues compared responses on the SWLS to responses on other existing measures of subjective well-being and related constructs such as affect intensity, happiness, and domain satisfaction. The findings confirmed their expectation that SWLS scores correlate highly with those measures that also elicit a judgement on subjective well-being and less so with measures that focus only on affect or self-esteem or other related but distinct notions. One piece of evidence in favour of the SWLS was that the scores of 53 elderly people from Illinois correlated well to the ratings this same population received in an extended interview about 'the extent to which they remained active and were oriented toward self-directed learning'(Diener et al., 1985, p. 73). The correlation, $r = 0.43$, was judged adequate by the standards of the discipline. Since 1985 the SWLS has continued to be scrutinised for its agreement with the growing data about subjective well-being. Individual judgements of life satisfaction have been checked against the reports of informants close to the subjects (Schneider & Schimmack, 2009). Proponents of the SWLS argue that it exhibits a plausible relationship with money, relationships, suicide, and satisfaction with various domains of life, such as work and living conditions.[8]

8. See Diener et al. (2008, pp. 74–93) for summary and references.

5.4.2. Construct Validation Is Good, in Theory

The first step to evaluating these practices is to capture their rationale. Neither philosophers nor scientists themselves have done that, so I propose to start with the following schema that summarises all the grounds on which a measure can be declared valid in this particular tradition:

> *Implicit Logic*: A measure M of a construct C is validated to the extent that M has been shown to behave in a way that respects three sources of evidence:
>
> 1. M is inspired by a plausible theory of C.
> 2. Subjects reveal M to track C through their questionnaire answering behaviour.
> 3. Other knowledge about C is consistent with variations in values of M across contexts.

The first condition captures the role of philosophising about the nature of C in the first stage of measure development. The second condition specifies the assumption behind factor analysis. It helps to reveal the structure of the construct as respondents see it (more on this in Chapter 6). The third acknowledges that scientists go beyond the merely internal analysis of the scale: a valid measure correlates with indicators that their background knowledge says it should and does not correlate with indicators that it should not.

There is much to be said on the foundations of this tradition. Philosophers of measurement roughly agree that goodness of a measure lies in its ability to mirror the behaviour of the target system. But how can such mirroring be established? Here there is no single story. Erik Angner (2009, 2011a) argues that there are at least two traditions—axiomatic and psychometric—corresponding respectively to economic and psychological approaches to the measurement of well-being.[9] In economics the key to the measurement of well-being is a representation relation between preferences and behaviour contained in axioms of the fundamental utility theory, while in psychology and the clinical

---

9. Angner (2015) argued more recently that the social indicators approach, motivated by an objective list theory of well-being, has yet a third measurement story to offer.

sciences the key is a valid questionnaire. The leading account about how mirroring is accomplished is known as the representational theory of measurement. According to it a measure is validated if there is a demonstrated homomorphism between an empirical relational structure (say an ordered series of rods for measurement of length) and a numerical relational structure (relations between real numbers). Various systems of axioms establish the conditions for this homomorphism.[10] However, the availability of a representational story for psychometric validation is far from obvious. Most scholars agree that the psychometric approach does not have the axiomatic basis characteristic of the representational approaches.[11] For example, there is no proof that an ordinal scale such as the SWLS can be treated as an interval scale, which nevertheless typically happens when it is used to rank, say, countries. In foundations of economics there are axiomatic structures to take us from preferences to utility functions and to choice behaviour but typically not so in psychology.[12]

My bet, though here I just state rather than defend it, is that the coherentism of Implicit Logic makes up for the lack of representational theorems. If a measure really behaves in accordance with background knowledge, then this alone is enough to secure its validity. Construct validation as described earlier conceives of measurement as part of theory development and validation as part of theory testing. On the original proposal formulated in the classic 1955 paper by Lee Cronbach and Paul Meehl, construct validation consists in testing the nomological network of hypotheses in the neighbourhood of the construct in question. To measure x, we need to know how x behaves in relation to other properties and processes that are systematically connected with x by law-like regularities (Cronbach & Meehl, 1955). Something like this view is still the consensus among scientists of well-being: 'To determine whether a measure is useful, one must conduct empirical tests that examine whether the measure behaves as

---

10. On general theory of measurement see Suppes (1998), Tal (2016), Cartwright and Bradburn (2011). The representational account is developed in Krantz et al. (1971).
11. I have in mind Wilson (2013), Angner (2009), Borsboom (2005, p. 86), and Michell (1999).
12. Though see Kahneman et al. (1997) for an example of a representational approach to happiness.

would be expected given the theory of the underlying construct' (Diener et al., 2008, p. 67).

This vision of measure validation is *prima facie* defensible. Its spirit is remarkably similar to the coherentist vision that characterises recent work on measurement of physical quantities.[13] The historians and philosophers of measurement emphasise that the outlines of the concept in question, be it temperature or time, and the procedure for detecting it, are settled not separately but iteratively, checking and correcting one against another. Similarly in our case, the initial philosophical judgement about the nature of happiness or quality of life is coordinated with other constraints such as the statistical features of the questionnaires and the background knowledge about behaviour, related indicators, and ratings of informants. The resulting measurement tools can be deemed valid to the extent that they accommodate all evidence.

Of course, scientists readily admit that validation is a continuous process, that it is never strictly speaking over, and that measures need to be revalidated for each new environment and population. Validation of the SWLS described earlier did not stop many sceptics from raising questions about the relation between life satisfaction and actual subjective well-being. As we have already seen, critics accused life satisfaction judgements to be ad hoc constructions that sway with arbitrary changes in the environment. These are the criticisms that Hausman (2015, p. 110) invokes against life satisfaction. He neglects to mention, however, the lengths to which psychologists have gone to check whether life satisfaction judgements are quite as fragile. It turns out that they are not, and today the SWLS continues to be popular partly because these judgements are more robust than its critics alleged (Lucas, 2013; Oishi et al., 2003).

The story of validation of the SWLS is typical. All the questionnaire-based measures of health-related quality of life, flourishing, and emotional state go through a similar process.[14] I recount this story in detail

---

13. Chang (2004), van Fraassen (2008), Tal (2013).
14. See Lyubomirsky and Lepper (1999)'s Subjective Happiness Scale, Diener, Wirtz, et al. (2010)'s SPANE and Flourishing Scale. All are examples of validation of well-being relevant measures using roughly the same methods.

to emphasise that measures of well-being are not selected haphazardly, not normally anyway, and it is thus not fair to criticise them by appeal to any particular apparent problem. Instead we need to criticise the whole package, and this is another part of my reply.

I suspect that Hausman would welcome attention to psychometric validation. But does this methodology offer a solution to his heterogeneity problem, which, to remind, stems from the fact that different goods, even when faithfully detected by well-being questionnaires, have different values for different people? Potentially yes. If a questionnaire really does agree with all of the relevant background knowledge as construct validation aspires to ensure, then the mere intuition that this questionnaire *could* go wrong in some individual case remains just that, an intuition. I imagine that believers in construct validation would reply to Hausman's worries in just this way—we will only worry about heterogeneity if the data indicate that our questionnaires do not behave as they should.

This is a fine response as far as it goes, and it is enough to establish that the sweeping sceptical verdict Hausman reaches is not warranted on the basis of reasons he provides. In actual fact construct validation does not live up to its great ambition to check questionnaires against *all* the relevant knowledge. This is my argument in the final chapter. For now I take stock.

5.5. MEASURABLE AFTER ALL

I have argued that well-being could be measurable if we focus on contextual rather than general well-being and if our measures behave in a way that coheres with all the available evidence. Does my case make a serious dent in Hausman's argument? Yes and no.

No, because I have said little to undermine Hausman's main contention that well-being in its most expansive sense is not measurable with the current (or possibly any) tools. Scholars of happiness may well conclude that I have yielded too much ground. Positive psychologists, whose whole enterprise is premised on marshalling scientific method to improve overall well-being for individuals and organisations, will

be particularly unimpressed by my concession. I intend this consequence. On my story the science of well-being need not strive to be
the science of all-things-considered well-being to justify its existence.
Instead it is on firmest ground when it concentrates on well-defined
kinds and contexts. If I wanted to know how to do better in life, well-
being science would not be my first destination. It may be informative
about specific components of well-being, such as Haybron-happiness,
healthy relationships, mental health, and so on. I may also learn useful
facts about major components of my well-being to the extent that I am
a member of a kind whose well-being is well understood by science. But
science is unlikely to speak more directly about my overall well-being
as a unique person that I am. For that I would instead go to someone
who truly knows me, who can judge my own personal well-being stew
in Hausman's sense—an old and thoughtful friend, a wise therapist, a
mentor, a trustworthy religious leader if I were a believer. None of this
justifies abandoning the pursuit of this science; it is just the realistic
interpretation of the available knowledge.

Also disappointed will be those for whom this general sense of
individual well-being is the most central and significant to human
life. To them my invocation of a different, contextual sense of well-
being will come across as lowering the bar in a way that makes the
concept lose its unifying force in human life. If such a redefinition
serves only the goal of making measurement possible, that seems
like putting the scientific cart before the philosophical horse. Again
this is intentional. Redefinitions happen. One of the lessons of recent
work on history and philosophy of measurement is that the theory
of the phenomenon and its measurement *co-evolve*. To quote Bas van
Fraassen (2008, p. 116): 'The questions *What counts as a measurement
of (physical quantity) X?* and *What is (that physical quantity) X?* cannot be answered independently of each other' (author's italics). To
apply this insight to our case, it is no good to decide ahead of time
from a philosopher's pedestal what well-being is and then declare that
no measure can do justice to this notion. The practicalities of measurement, the need for common reliable standards that enable comparisons and scientific communication, should all naturally inform
the shape of concepts we posit. This mutual correction of scientific

requirements and philosophical constraints (plus political and cultural ones) is the story of science. If well-being is to play a useful role in life of today's bureaucracies, which live by numbers—and this train appears to be unstoppable now—well-being may have to be *made* measurable even if it was not initially.

Chapter 6

Psychometrics as
Theory Avoidance

I have argued that there are measurable properties worth calling well-being. Their measurability is secured by the existence of generalisations connecting component goods (e.g., social support) with the contextual well-being of members of the kind in question (e.g., caretakers of the chronically ill), as defined by the relevant mid-level theories (e.g., well-being as being protected from strains of caring). When scientists claim to measure well-being in this sense, the validity of these measures is established usually through construct validation, a procedure whose logic I characterised in Chapter 5. I painted a positive picture on which this procedure checks that a given questionnaire exhibits correlations that agree with the background knowledge about the construct in question and to this extent construct validation guarantees epistemic consilience. Our measures of well-being are only invalid if the rest of our knowledge is invalid too.

But it is one thing to describe construct validation in theory and another to check whether it actually works in practice. This is the task of this chapter, and this time my message is less optimistic. In the admirable logic of construct validation, much rides on how background knowledge is defined and how its deployment is implemented. It is easy to say that a good measure of a phenomenon should be based on and responsive to our best theory of this phenomenon. The harder question is what counts as such a theory. I have argued throughout the book that the science of well-being often lacks explicit theories against which its measures should be judged and that philosophers have not provided them as much as they should have. Construct validation is a case in point—it proceeds not so much in the absence of theory (for that it is impossible) as in wilful ignorance of some theoretical knowledge even when it *is* available. Psychometric validation is selective about which knowledge it admits, and in this sense it is *theory avoidant*. I frame this book as a meditation on what philosophy science needs and what philosophy it can ignore. But some practices of psychometric validation strike me as problematic insofar as they aim to validate measures of well-being while paying only scant attention to what well-being is in any given case and this is exactly the wrong philosophy to avoid.

More precisely construct validation as practiced avoids relevant theory by adopting several stances that I group under the label *evidential subjectivism*. Evidential subjectivism is a bet that not only is the object of measurement always a psychological state but, more importantly, the evidence used to validate a measure of this state must itself feed mainly from the reports or behaviour of the relevant subjects in relation to this measure. Whatever philosophical evidence we might have about the nature of the psychological attribute in question or its normative significance for well-being is at worst ignored and at best discounted. This amounts to theory avoidance because the substantive questions about the nature of the psychological states relevant to well-being are reduced to observations of the behaviour of respondents to questionnaires. In Chapter 4 we saw that objectivity of well-being research sometimes requires facing up and addressing openly and deliberatively big questions about values. But it is hard to find room for this in the psychometric practices, which are largely technical exercises. If the goal is valid measures of well-being, we must do better.

6.1. WHAT CONSTRUCT VALIDATION LEAVES OUT

To begin to see the scope of the problem recall my implicit logic from Chapter 5—only this time we shall look at it critically:

> *Implicit Logic*: A measure M of a construct C is validated to the extent that M behaves in a way that respects three sources of evidence:

1. M is inspired by a plausible theory of C.
2. Subjects reveal M to track C through their questionnaire answering behaviour.
3. Other knowledge about C is consistent with variations in values of M across contexts.

None of the conditions as they are currently implemented are strong enough to ensure validity. Condition 1 apparently opens the door to philosophical examination of the scope and the meaning of, in our case, well-being—a good step—how else should one start the measurement process? But in practice there are no strong criteria for what makes a conception of C plausible, how elaborate it should be, how systematically the alternative conceptions should be considered and evaluated. What well-being is, what it encompasses, and what falls outside its scope are big questions that consume many person-hours of philosophers. But for scientists who are not trained to philosophise and whose identity depends on not being philosophers, this task is far less compelling. Indeed the philosophical heart of Condition 1 often enough is replaced by an informal report of folk views or an unsystematic literature review. Instead of examining the nature of well-being of the relevant kind by building at least in outlines a mid-level theory of it, the temptation is to canvass how this concept is understood by the relevant population and be done. I document this trend of substituting reports of subjects for systematic theorising in the next section. It is undoubtedly a good idea to canvass folk views on well-being, but these views—mostly a collection of platitudes—are unlikely to serve as sufficient constraints for fulfilling Condition 1.

A weakly implemented Condition 1 affects the data that go into Condition 2. In a joint paper with Dan Haybron we note this failure in a commonly used and conventionally validated scale of happiness called the PANAS (Positive and Negative Affect Scale; Alexandrova & Haybron, 2016). The PANAS invites ratings on whether subjects feel enthusiastic, interested, excited, strong, alert, proud, active, determined, attentive and inspired, and so on (Watson et al., 1988). My coauthor is unimpressed:

> note that absent from this list are cheerfulness, joy, laughter, sadness, depression, tranquillity, anxiety, stress, weariness—emotions

that are intuitively far more central to a happy psychological state, and to well-being. This is because the authors of PANAS arrived at the list of items by testing a long list of English mood terms and paring it down via factor analysis, so that a longer list would not yield appreciably different results. Such a procedure allows investigators to avoid hard theoretical questions about which taxonomy of emotional states to employ, or which states are most relevant to well-being. But for the same reason, there is little reason to expect such a method to yield a sound measure of well-being, or even of emotional well-being. Rather, what is being assessed, roughly, is the number of English mood terms that apply to the respondent—or rather, the number of terms from a list of words that survived factor analysis. But, first, this leaves the measure prey to the vagaries of common English usage and folk psychology—potentially important emotional phenomena may not be prominent in the vocabulary of a given language, or may not be correctly classified as emotional, and so may be omitted from the measure. Of particular concern here are relatively diffuse background states—anxiety, stress, peace of mind (not on the list)—that are quite important for well-being yet easily overlooked, resulting in a kind of "streetlight" problem where we end up looking where the light is best, rather than where the keys are.

Second, some states are presumably more important for well-being than others; feelings of serenity or joy (not on the list) probably count for more than feeling "attentive" or "alert" (on the list), and indeed some of the PANAS items might barely deserve inclusion at all, if our interest is in assessing well-being. Yet a term like "attentive" might exhibit quite distinctive correlations, and thus make it on the list, while other more salient terms are left by the wayside. (Alexandrova & Haybron, 2016, p. 1106)

This dramatic failure of PANAS to appreciate the complexity of what it means to feel happy stems perhaps from a desire to stick to easily available questionnaire data and familiar tools of factor analysis. These come easier to scientists than the sort of philosophical work that led Haybron to formulate his emotional state theory and to explain its significance for well-being.

And it is no good to hope that implementing Condition 3—the coherentist heart of the whole exercise—will correct the mistakes made in Conditions 1 and 2. Condition 3 invites scientists to study correlations between the measure they are validating and other related measures and factors with which they expect their construct to correlate or countercorrelate. But again correlations on their own, without an explicit theory, are not discriminatory enough. If a measure correlates in expected ways with suicide rates and self-harming, health, smiling, cortisol levels, and so on, this is evidence that this measure is plausible. But in addition to plausibility we need evidence that this measure is better than another plausible measure. As we argue, this is where correlations often give out:

> take a long list of variables that seem like they might be related to well-being—money, relationships, health, education, work, etc. Imagine two measures, A and B, each of which correlates substantially with nearly all of these variables, while also differing greatly in what those correlations are. One suggests that relationships are more strongly related to well-being than money, while the other has the reverse implication, and so forth. It seems entirely possible that both measures could reasonably be deemed to exhibit "plausible correlations," and generally pass as valid measures of well-being. It is also possible that one of those measures is *in fact* valid, while the other is not: A gets the correlations essentially right, while B gets them wrong. (Alexandrova & Haybron, 2016, p. 1104)

We go on to show that this is precisely the shape of the conflict between metrics of life evaluation such as the SWLS and affect measures. Empirical evidence increasingly indicates that the former track material circumstances, while the latter track 'psychosocial prosperity' (which includes being treated with respect, having people to count on, doing something you are good at, having autonomy and other good things; Diener, Ng, et al., 2010). Which correlation is more significant?

The existence of such questions shows that construct validation cannot be just about balancing our web of belief by checking which measure accords maximally with our existing knowledge. There will remain genuine normative questions about whether 'psychosocial prosperity' is more important than a material one. There will also be questions about how important life satisfaction is for well-being. Previously Haybron argued that to be satisfied with life has to do with the values one endorses such as gratefulness, modesty, determination. Life satisfaction reflects 'one's stance toward one's life' (Haybron, 2008, p. 89). But the stance I adopt toward my life is one thing, he maintains, the state of my life—my emotions and my daily quality of life—is another. But there are no provisions in the methodology of psychometric validation to take this distinctly philosophical considerations on board. These questions cannot be resolved by checking more correlations. Rather they need an explicit deliberation of what counts as well-being to a given community, and there are no resources in Implicit Logic for such deliberation.

The solution must be to enrich Implicit Logic by blocking the current status quo of theory avoidance and by making room for philosophy and values. First, it is worth going deeper into the sort of disciplinary conventions that enable theory avoidance.

6.2. SUBJECTIVISM IN TEXTBOOKS

In formulating Implicit Logic in Chapter 5, I took cue from rules enshrined in textbooks. I want to return to one in more detail. *Measurement in Medicine*, published in 2011 by a University of Amsterdam clinimetrician Henrica De Vet and her colleagues, offers a comprehensive and clear guide to developing measurement instruments for all medical and health fields (De Vet et al., 2011). There is no dedicated textbook on measurement of well-being yet, but the procedures described by De Vet and her coauthors are mostly the ones followed by psychologists and social scientists of well-being.

Consider one kind of subjectivism. Speaking of Health Related Quality of Life (HRQL) De Vet and coauthors (2011, p. 11) note: 'HRQL . . . can only be assessed by PROs because they concern the patient's opinion and appraisal of his or her current health status'. 'PROs', to remind, stands for Patient-Reported Outcomes, a term in medical research that denotes patients' own assessment of their state. This sort of subjectivism—that the target phenomenon is a mental state—characterises a big portion of the sciences of well-being. Development economists typically depart from this approach, but here I neither endorse nor complain against this focus. PROs and subjective well-being in general are subjective by design in the sense that they are representations of people's own evaluation of their lives, rather than their objective quality of life or their sets of capabilities. Subjectivism in this sense is a choice scholars make for various reasons—because this is the perceived purview of psychology as a discipline, or as a reaction to past exclusion of patient perspectives, or for some other reason.

My focus is not this perfectly explicit commitment to study the subjective aspects of well-being. Rather I wish to observe that only a certain type of evidence is allowed to bear on whether or not a given questionnaire asks the right questions about this subjective state. The methodology for validating these questionnaires is geared toward discounting philosophical considerations about the nature of the subjective state in question in favour of evidence derived mostly from reports or behaviour of subjects. So this methodology forces a very specific kind of *evidential* subjectivism.

To settle on the questionnaire items researchers are usually directed to the existing 'item banks' such as the National Institute of Health's PROMIS (Patient Reported Outcome Measurement Information System), which already contains pretested scales for use in any clinical or research setting. But if researchers insist on developing their own questionnaire items, after formulating the items and settling on the scoring methods the first task is to test whether these items are well-behaved from the psychometric points of view. This is the start of pilot testing. The textbook is clear: 'only the target population can judge comprehensibility, relevance and completeness [of the questionnaire]' (De Vet et al., 2011, p. 58).

It is worth pausing on this requirement. It is perhaps uncontroversial that only the people themselves can judge whether the questions they are asked are comprehensible to them. But relevance and completeness are different. When we say that only the people whose subjective state is in question can judge whether or not the questions asked about their subjective state are relevant and complete, we are taking a *doubly* subjectivist stance that is far more controversial than the initial choice to focus on the subjective states. Suppose the goal was to measure my happiness. It is certainly appropriate to check that my understanding of the related concepts (contentment, peace, elation, engagement, etc.) matches how the measurer understands these concepts. But it does not follow that I am the only authority on what happiness is. There are better and worse theories of happiness as we have seen, and this plain fact appears to be denied by evidential subjectivism, which in turn appears to be written into the very procedure of measure validation. Here is more textbook advice along the same lines: 'the importance of the items has to be judged by the patients in order to decide which items should be retained in the instrument' (De Vet et al., 2011, p. 66). Also,

> the most important items should all be represented. This implies that for the decision with regard which items should be included we need a rating of their importance. These ratings of importance can be obtained from focus groups or interviews with patients. (p. 70)

More evidence of this attitude can be seen in discussions of content validity, a type of validity distinct from construct and other validities.[1]

6.2.1. Subjectivism and Content Validity

Content validity is a slippery concept but on most definitions it assesses whether the measure is *about* the right thing, that is, whether the content of the measurement instrument matches the content of the construct measured (De Vet et al., 2011, p. 154). 'Adequate representation' of the construct by the instrument is another way used

---

1. Criterion validity, which compares a measurement instrument in question to the golden standard, is not used in the science of well-being as it is thought that the golden standard in this area, or any other area where patient reports are used, does not exist (De Vet et al., 2011, p. 161).

to describe this match (p. 155). The requirement seems to be that for any well-being questionnaire content validity obtains to the extent that its items capture well-being given the commonly agreed content of this concept.

Content validity is rightly controversial. If it requires that a measure fully captures the agreed content of a construct, then it appears to exclude indirect measures. What if the best measure of an attribute captures this attribute via an indicator with which the attribute is reliably correlated, rather than trying to represent the attribute directly? Mid-upper arm circumference is a fine measure of malnutrition without capturing the content of malnutrition. So here is a case in which construct and content validity appear to be in tension. If, however, such indirect measures are compatible with content validity, then it becomes hard to differentiate content from construct validity. This is the sort of argument that has led psychometricians to treat construct validity as primary. Perhaps, some have argued, content considerations should inform the first stage of measure development in which the scope of the concept in question should be delineated fully and with care, but after that construct validity should prevail. [2] On the other hand, content validity is currently seeing a revival as measures of PROs in healthcare are increasingly judged on their acceptability and penetrability to the patients who fill in these questionnaires.[3] Perhaps the role of content validity is political—to ensure that measurement instruments look sensible to their audiences.

Let me sidestep these debates. Content validity is used rightly or wrongly to assess well-being questionnaires. My question here is whether the methods of its assessment follow evidential subjectivism we have seen in the earlier steps of psychometric validation. If content validity is also problematic for other reasons, that is orthogonal to my focus. So how is content validity evaluated?

Content validity assessment is in general a qualitative exercise. De Vet's textbook reflects the common disciplinary standards in proposing

---

2. For a history of these debates, an interpretation and a defense of content validity see Sireci (1998).
3. McClimans and Browne (2011) survey this trend.

that content validity be judged on the basis of 'relevance' and 'comprehensiveness' of its items. Relevance makes sure that of all the items none are superfluous given the nature of the population to which the measure is to be applied. (For example, cis men should not be asked about whether they've had a hysterectomy.) Comprehensiveness, on the other hand, is a check that all the items that should have been included are included. There is a set procedure, indeed a checklist, for assessing relevance and comprehensiveness.[4] But chief among these procedures and the most interesting for us is the recommendation to use expert panels: 'For all measurement instruments, it is important that content validity should be assessed by experts in the relevant field of medicine' (De Vet et al., 2011, p. 157).

It is hard to argue with this recommendation. A panel of experts of child development is indeed a good, if not the best, authority on whether a given measure of child well-being captures what it is supposed to capture. Similarly, experts in radiology are best positioned to evaluate uses of MRIs for diagnostics and so on.

But note what happens when the attribute in question is well-being or any other of the PROs. On the face of it a panel of experts here is perfectly conceivable: it could include social workers, psychotherapists, other therapists, psychologists, economists, sociologists, anthropologists, not to mention philosophers, all of whom have a great deal to say about the nature of human flourishing and its social, cultural, and economic preconditions. But that is not what happens:

> For patient-reported outcomes (PROs), patients and, particularly representatives of the target population, are the experts. They are the most appropriate assessors of the relevance of the items in the questionnaire, and they can also indicate whether important items or aspects are missing. (De Vet et al., 2011, p. 157)

Evidential subjectivism here is out in the open. Subject agreement with the measure is treated as necessary and sufficient for content validity.

4. http://www.cosmin.nl/cosmin-checklist_8_0.html. See also Mokkink et al. (2010) for a proposal on how to standardise the procedure for all steps of measure validation.

6.2.2. Subjectivism in Factor Analysis

A different way of restricting evidence is effected through the use of statistical techniques, for example, the aforementioned factor analysis. In Chapter 5, we came across its use to discover and confirm the structure of target constructs.

On the basis of factor analysis, researchers know, for example, that subjective well-being breaks up into several distinct components: positive affect, absence of negative affect, and life satisfaction. They are especially proud of having demonstrated 'empirically', that is by factor analysis, that presence of positive affect is not the same as absence of negative affect; rather the two are separate independent factors.[5]

In case of the SWLS (which to remind is Satisfaction with Life Scale), the factor loadings of items ranged from .61 to .84 and were deemed to be good evidence that life satisfaction is a unified phenomenon on its own. If the factor loadings were lower, that could have been taken as evidence that the different items of the SWLS are not measuring one thing. Depending on the prior expectations about the dimensionality of the construct, such results would argue against the inclusion of these items into the questionnaire. Similarly, psychologists have used factor analysis to show that subjective well-being with its three components is distinct from what some call flourishing, or psychological well-being (PWB). The latter is a concept arising out of humanistic traditions in psychology and identifies well-being with perceived autonomy, mastery, connectedness, and growth. It may come naturally to philosophers to treat it as distinct from flourishing, but for psychologists this distinctness is always an empirical question and factor analysis is the typical tool for the demonstration of this fact. For example, Corey Keyes and his coauthors argue:

> Using a national sample of US adults . . . we confirm the hypothesis that SWB and PWB represent related but distinct conceptions of well-being. The data indicate that the best fitting model is the one that posits two correlated latent constructs, namely SWB and

5. Lucas et al. (1996), Arthaud-Day et al. (2005), Diener and Emmons (1985).

PWB, rather than two orthogonal factors (or one general factor). Thus although these latent constructs are highly correlated, each retains its uniqueness as a distinct facet of overall well-being. (Keyes et al., 2002, p. 1017)

The difference in disciplinary conventions is particularly striking here. While philosophers would approach the question as conceptual (Is the concept of well-being unified?) or metaphysical (Is well-being unified?), for psychologists the question is whether the answers to questionnaires reveal statistical structures that allow us to postulate different factors 'driving' well-being. Such diversity in methods is entirely fine. And no doubt valuable knowledge can be gained from factor analysis—namely knowledge about patterns of questionnaire responses. But is this knowledge critical for validity? It is certainly treated as such. Once the factors have been identified by factor analysis, it becomes necessary to fit the measures to the model confirmed by this method. If you wish to measure well-being, then you should aim to have items in your questionnaire that represent the core elements of well-being as identified by factor analysis.

This too is a kind of evidential subjectivism. The assumption appears to be that only an item that subjects' behaviour reveals to be significant in their understanding of well-being should be part of a measure of well-being.

So we have two kinds of evidential subjectivism: questionnaire items are valid only if they have the right psychometric properties and questionnaire items are valid only if the subjects say so. We can call the first *behavioural* subjectivism since here subjects are not asked whether they think a given item asks the right question. Instead this is inferred from their questionnaire answering behaviour via techniques such as factor analysis. The second, a more traditional approach, can be called *conversational* subjectivism, since it requires actually talking to the subjects.[6]

6. Which mode of subjectivism is adopted in practice seems to depend on whether the construct is treated on a reflective or a formative model. On the reflective model the construct is the latent variable, which is picked out by its manifestations (e.g., anxiety is reflected by panic, restlessness, etc.). On the formative model the construct is picked out by its causes rather than its consequences (e.g., quality of life is formed by satisfaction

6.2.3. Against Evidential Subjectivism

Let us distinguish evidential subjectivism from other forms of subjectivism. In moral philosophy, subjectivism about well-being has a more or less stable meaning. Consider a recent characterisation: 'subjectivism about well-being holds that φ is intrinsically good for *x* if and only if, and to the extent that, φ is valued, under the proper conditions, by *x*' (Dorsey, 2012, p. 407, emphasis removed). Subjectivists then argue about the nature of these proper conditions and about what valuing should consist in.

The sort of subjectivism that underlies psychometric methodology is dramatically different. The following is a formulation following this model:

> *Evidential subjectivism*: φ should be accepted as a component of *x*'s well-being only if φ is a self-report of a factor shown to capture the data of well-being questionnaires completed by subjects relevantly similar to *x* (behavioural subjectivism) *or* φ is a self-report of a factor systematically claimed to be valued by subjects relevantly similar to *x* in a well-designed interview or survey (conversational subjectivism).

In this definition I tried to pack in several features of evidential subjectivism. First, I distinguish between behavioural and conversational subjectivism with a disjunction. This disjunction is not exclusive. Questionnaire items can be and often are subjected to both requirements—to have good psychometric properties *and* to be acceptable to subjects. Second, there is no requirement that *x* herself completes the questionnaires or undergoes the interview; rather the items of the questionnaire need to be validated in a representative sample

with domains of life). The formative model does not depend on statistical item reduction as much as the reflective model does (De Vet, 2011, p. 71). But there is no agreement as to whether well-being and the related constructs should be treaded on formative or reflective models; indeed there are examples of both approaches (Sirgy, 2002, 2012). Life satisfaction and happiness are usually approached on the reflective model and quality of life on the formative one.

after which it is permissible to generalise beyond this sample. I capture this with a phrase 'subjects relevantly similar to *x*'. Finally, the 'only
if' formulation specifies that a questionnaire passing this test is only a
necessary, not a sufficient, condition for its validity. This is because, in
accordance with Condition 3 of Implicit Logic, scientists go beyond
evidential subjectivism, making sure that the questionnaires also correlate appropriately with behavioural and socioeconomic data and other
plausible questionnaires.

Of course, scientists go beyond evidential subjectivism in other
ways too. In conversation, practitioners routinely recognise that good
psychometric properties and endorsement by subjects are not sufficient
for validity. It is accepted that the questionnaire items subjected to statistical analysis also have to be well grounded 'theoretically'. Only after
the initial bit of philosophy, evidential subjectivism kicks in to determine which of the theoretically grounded questions are the right ones
to include. What is clear, however, is that the philosophical conjectures
about the nature of target constructs are easily overridden. Although
there is nothing specifically that forbids Condition 1 of Implicit Logic
from trumping Conditions 2 and 3, it is in effect very hard for philosophical considerations to get through. Psychometrics wins just in virtue of
being performed *after* the theory-based choice of questions.

This is the sense in which evidential subjectivism enables theory
avoidance. Systematic philosophical accounts of well-being and related
constructs offer the only opportunity to theorise normatively, that is, to
ask what is happiness, or life satisfaction, or flourishing, such that they
can play the role that these thick concepts play in people's lives. And yet
this normative test is not part of measure validation, except perhaps as
an initial inspiration. This is insufficient if the outcome is supposed to
be a measure of a property denoted by a thick concept. Part of measure
validation should be whether the measure captures a construct that is
worth caring about.

I contrasted evidential subjectivism to the subjectivism familiar to
philosophers on purpose to bring out the fact that the standard justifications used for the latter will not work for the former. Subjectivism
in Dorsey's formulation is usually put forward as the best compromise
between two intuitions: first, that something can only be good for an

agent if that agent finds it compelling or attractive, and second, that people can and do regularly make spectacularly bad judgements about what is good for them (see Appendix A for references). Evidential subjectivism that we see in psychometrics must have very different roots. It shows virtually no attempt to mitigate the poor judgements that philosophers so worry about, and there is little effort in this exercise to find out what people genuinely value and why.[7]

I can think of three possible justifications for evidential subjectivism. The first is the most straightforward one: the users of psychometric validation accept it as the correct substantive theory of well-being. This possibility can be dismissed quickly. It is unlikely that all the diverse users of the techniques in question agree on a single claim about the nature of well-being, let alone one that is so specific. There are far more plausible versions of subjectivism if that is what we want.

My second guess imputes to the adherents of evidential subjectivism a methodological stance rather than a commitment about the nature of well-being. Perhaps this stance expresses the only conditions under which psychologists can know about well-being, if they can know anything about well-being at all. Whatever well-being is, if it is epistemically accessible, it would be accessible by the methods endorsed by evidential subjectivism. This option presents itself as the modest stance. The tools of science are only suitable to finding out about self-reports of mental states (or their manifestations observable via answers to questionnaires) and nothing else. Well-being could well be a lot more, but then it could not be an object of science. Before evaluating this stance I present another possibility.

My third guess interprets evidential subjectivism as a political stance: out of respect to the subject's autonomy, the science of well-being should not attempt to go beyond their judgements of their own well-being. On this interpretation, evidential subjectivism, though not without normative presuppositions, is saddled with the least controversial

---

7. I say 'virtually' because life satisfaction proponents might argue that they do make an honest attempt to put subjects in the right frame of mind when they invite them to judge their satisfaction with life. The questionnaire items are prefaced with 'All things considered . . .' or 'Taking all the relevant things into consideration . . .'. But this hardly counts by the lights of philosophical subjectivism.

and the most neutral assumptions. Science does not have the authority to go any further.

Neither of the last two interpretations are defensible. Evidential subjectivism is neither neutral nor modest.

I start with the political version. Those who believe that values should not encroach on scientific methods will reject it outright, but I am not one of them. To the extent that representing the perspective of the subject is one goal of the science of well-being, a political justification of a measurement procedure can be appropriate. Indeed this justification describes well the intent behind the inclusion of patient reports in evaluation of clinical practice – patients need a voice and these measures provide it. But notice that this political justification has to apply to both parts of evidential subjectivism—behavioural and conversational—but it applies if at all to one. I can see how conversational subjectivism is a bone fide attempt to defer to and to respect the views of the subjects, though even here we might worry that a structured interview or a preset questionnaire does not accomplish the goal of genuinely learning our subjects' perspective.[8] But behavioural subjectivism—where the subjects' concept of well-being is inferred from the statistical patterns of their questionnaire answers—does no such thing.

The second problem with the political defense is that respecting people is tricky, and sometimes inferring their views from their behaviour or even their comments on questionnaires is not respectful. More than any other discipline psychology can be credited with discovering

8. McClimans (2010) notes that communicating with subjects only via prepackaged questionnaires is not a proper conversation because it does not ask 'genuine questions', that is, questions that really allow the subjects to contribute on their own terms, not just on the terms of the clinician posing the questions. McClimans would argue that a proper respect for subjects requires something closer to an open dialogue. But it is important not to issue scientists advice verging on perfection. Diener and his colleagues did after all check the SWLS against a lengthy structured interview and De Vet's textbook goes out of its way to urge researchers to talk and *hear* the patients' point of view. This could be done by leaving a blank page at the end of the questionnaire and asking subjects to comment on whether they have anything else to add. There are other techniques to enable a more serious engagement with the views of those whose well-being is being studied. They are not always followed, and perhaps this is McClimans's complaint, but the idea of what I have called conversational subjectivism exists, and researchers do make the bona fide attempts to stay rooted in the priorities and views of their subjects.

all the ways in which humans can fail to know their inner state; posi-tive or negative; past, present, or future.[9] This is *affective ignorance*, to use Haybron's term. It is ironic then that this very discipline defers so much to our own assessments—behavioural or conversational—of what questions can best detect our true inner state. It seems particu-larly questionable to justify this stance on grounds of respect. Would it not be more respectful to adjust the methods to facts of life? Would it not indeed be *dis*respectful to adhere to evidential subjectivism while also knowing that our judgements can be seriously mistaken? Perhaps in reply psychologists would appeal to the fact that they do not know for sure when these mistakes happen and when they do not; and hence it is safer to defer to people themselves on the off chance that we discount their judgements when we should not have. But that is just implausible. Parents know very well when to discount the judgement of the child who claims not to be sleepy when it is in fact past his bedtime. A good friend also knows when to take seriously her friend's admission of distress—so do therapists, so do social workers, and so on. Respect is compat-ible and indeed requires sometimes discounting some judgements of those to whom we owe respect. [10] True, sometimes we should defer to judgements even though we have good reasons to suspect them to be flawed. Perhaps there are circumstances when science may simply lack the authority to substitute another judgement. Still it is hard to see how evidential subjectivism can be a blank check. If the nature of well-being raises fundamental conflicts of values (What matters more: happiness or satisfaction?), it is more respectful to the users of these measures to put these debates out in the open and adopt measures that pass minimal deliberation as I urged in Chapter 4.

Even without invoking affective ignorance we could raise ques-tions about the practice of asking people 'Is this a good question about your well-being?' as conversational subjectivism does. 'Good

9. For popular summaries of the empirical evidence for ignorance of our inner states see Wilson (2009) and Gilbert (2009). For philosophical significance of these findings see Haybron (2007), Schwitzgebel (2008).

10. We make a similar argument against mindless applications of cost-benefit analysis, which can end up oppressing people all the while pretending to avoid making paternal-istic judgements about their well-being (Haybron & Alexandrova, 2013).

for what? It depends on what you will do with this information' would be my reply. None of my criticisms imply that people should not be consulted or that their conceptions of well-being should never be inferred from their behaviour. Rather the information obtained in such a way should not be left unexamined and should not automatically trump other evidence.

How about the methodological defence of evidential subjectivism? To remind, this version takes subjectivism to be a constitutive rule of a psychological approach to well-being. I worry that it takes the existing methods of a particular corner of science and reifies them into the only methods this science could possibly adhere it. This is not unheard of. Economists Faruk Gul and Wolfgang Pesendorfer (2008) recently tried to make a similar case for 'mindless economics'. By definition, they claim, economics deals only with behaviour not its psychological causes and not its normative significance. But such foot-stamping seems parochial even to mainstream economists (Caplin & Schotter, 2008). Methodology is not a matter of definition. We have seen that the science of well-being boasts a great diversity of approaches to well-being, and not all of these approaches are committed to evidential subjectivism. Is there something about psychology and clinical sciences that makes evidential subjectivism about well-being inescapable? I do not see what.

There is no justification for evidential subjectivism, but I could venture an explanation for its persistence. Evidential subjectivism reigns because of disciplinary conventions and the operationalist heritage in psychology. Psychometricians feel averse to acting like philosophers, that is, to theorising about the nature of well-being in a way that breaks away from the data collected by questionnaires. They prefer to stay grounded in their subjects' behaviour. In this sense evidential subjectivism is a modest stance. But it is less modest once we see its political context. Historians of psychological sciences have long noted the convenience of the methods of these sciences to the political order in which they arose and endure. Psy scientists, to use Nikolas Roses's (1990, 1998) term that covers psychologists, psychometricians, psychotherapists, and even psychoanalysts, have long played a crucial role in the management of individuals in liberal democracies. Their authority as advisors depends on their adoption of technical methods for handling questions that were not previously within the domain of science—what it means to be normal, intelligent, well-adjusted, and so on. As Rose argues, this is how moral or prudential questions are turned into psychological ones. Similarly in our case, in undertaking the validation of measures of well-being, psychometrics puts itself forward as the arbiter of questions that are properly moral and political. But in the hands of psychometricians these questions become technical, a matter of calculation and narrow expert judgement. Appeal to the subjects behaviour or their reports is a standard move. It makes validation procedures seemingly democratic and grounded in facts—and evidence-based too, so very convenient.

Far from being modest and safe, this avoidance of philosophy and its replacement with a technical exercise in construct validation is epistemically wrong and morally dangerous. Before I comment on what is to be done, I summarise my argument and distinguish it from other criticisms of psychometrics.

6.3. OTHER COMPLAINTS ABOUT PSYCHOMETRICS

To take stock, Haybron and I have argued that while construct validation is sound in theory, in practice it is too selective about what background knowledge is considered. In the initial stage when questionnaires are developed, such selectivity means that the meaning and scope of concepts are not properly examined in the light of the best available philosophical theories. Measures of happiness such as the PANAS end up being based on dictionary lists of positive mood terms instead of plausible and systematic theories of the relevant emotional states. In the later stage of validation, when psychometricians examine correlations between the measure under consideration and other relevant indicators, the mere correlations or lack of them are not enough because we need normative theory about the nature of well-being or related concepts to decide which correlations matter and which do not. We called these problems theory avoidance. Haybron coined 'correlation-mongering' as a less neutral but very apt term. I have argued that underlying these ills is a deeper commitment to evidential subjectivism—a position that restricts evidence about the nature of constructs to reports of subjects and to examination of the statistical structure of their behaviour as they answer questionnaires. Aspects of these practices may be defensible but as a whole they turn validation into a narrow and technocratic exercise.

Complaints about lack of theory in psychometrics are not new, going back at least to Stephen Jay Gould's (1981) attack on the use factor analysis in IQ testing in *Mismeasure of Man*. It is worth distinguishing my worries from other recent ones.

There are theorists of validity who criticise the operationalist foundations of construct validation. Psychologist-philosopher Denny Borsboom (2005) believes that robust causal relations rather than correlations should ground construct validation. He urges that a measure is valid if and only if the construct exists and causally produces variations in the measurement outcomes. This requires a realist commitment to unobservable mechanisms for which correlations can serve only as evidence. By sticking to a resolutely antirealist metaphysics the psychometric approach wrongly outsources to statistics what is essentially a theoretical problem: What must well-being be like, as a causal system, for questionnaires to detect it?

I am sympathetic to the sentiment, but in general sceptical whenever any specific metaphysics, realist or otherwise, is claimed to be essential for the practice of science.[11] I wish to criticise evidential subjectivism while retaining agnosticism about such deep foundations.

But perhaps Borsboom's complaint is that when statistical correlations are praised above all for construct validation this takes away from understanding the process of measurement as a system of interacting parts—the researcher, the questionnaire, the subject, the testing environment. A measurement procedure is trustworthy to the extent that we are able to represent more or less accurately the interaction between the measuring agent and the physical system, teaches van Fraassen (2008, Chapters 7–8). On this criterion, argues Leah

---

11. Hood (2013) presents a good compromise on the question of realism in psychometrics.

McClimans (2017) well-being measures are invalid even if they have good psychometric properties. Lack of theory about how subjects react when well-being questionnaires are administered means that researchers are unable to interpret when a measured change is due to a change in the person's values and when it is due to a change in her living conditions. This stark conclusion seems to me to be too strong. Van Fraassen, Borsboom, and McClimans are correct to require a theory of the measurement process, but McClimans exaggerates the extent of our ignorance about it. Psychologists do worry about what they call *construct representation*—a theoretical account of the process of measurement—and do develop representations of the process of judgement of subjective well-being.[12]

My objection to evidential subjectivism is different. When subjects' behaviour and evaluation of the questionnaires are used as nonnegotiable constraints on these questionnaires' validity, there is no room for a normative discussion. Are subjects right in saying, or in behaving as if, a given question is a good question to ask about well-being? Do they have a plausible notion of the state in question? Perhaps there is a better one they did not consider. Perhaps, in keeping with affective ignorance, there are aspects of our emotional state that easily slip our attention and awareness—peacefulness, stress, and anxiety are Haybron's examples—and yet there are strong normative reasons to include them in our definition of happiness or other psychological states relevant to well-being. Happiness would not be important for well-being if it did not encompass these, and a measure of happiness that is not sensitive to these is seriously compromised. Similar normative concerns can be raised about measures of related psychological states such as life satisfaction and sense of flourishing.

This normative focus—whether the states and attitudes being measured are conceived in a way that makes them sufficiently relevant for well-being (and which one)—is distinctive of good philosophical theorising. To the extent that evidential subjectivism replaces this sort of theorising, it is a bad constraint to adopt.

---

12. Strauss and Smith (2009), Kahneman and Riis (2005), Kim-Prieto et al. (2005).

6.4. WHAT IS TO BE DONE?

I have argued that Implicit Logic is valid but poorly implemented. The following is a new formulation of the three conditions that brings the problems I discussed into the open:

> *A Better Implicit Logic*: A measure M of a construct C can be considered validated to the extent that M behaves in a way that respects three sources of evidence:
>
> 1. M is inspired by a plausible theory of C. This theory should be articulated as fully as possible and defended against alternatives.
> 2. M is shown to track C as C is understood and endorsed by the subjects to whom C is applied.
> 3. Other knowledge about C is consistent with variations in values of M across contexts. This knowledge should encompass normative significance of C, including moral and political context of the use of C.

The coherentist spirit is preserved, but I revised the three conditions as follows: Condition 1 is now strengthened with a requirement to start the process of construct delineation with a systematic exercise in analysis of the concept C represents (this is my cheer for the lately much maligned conceptual analysis). Condition 2 is a revision of evidential subjectivism—I hope to preserve the spirit of accommodating the perspective of the subjects without blindly deferring to their unconsidered views as is the case currently. Condition 3 now explicitly accommodates the possibility of normative knowledge. My intention is to beef up the stock of knowledge that bears on construct validation. Not only correlations with existing measures and objective indicators should count but also knowledge, in our case, about the nature of well-being. In Chapter 5 I argued that we should be most optimistic about measurement of contextual rather than general well-being. If I get my way, claims about contextual well-being will be increasingly encoded in mid-level theories of it—accounts of well-being or well-being relevant states grounded sufficiently in a practical context of their potential use. For the study of subjective well-being that I focused on in this chapter these would be theories of happiness, life satisfaction, sense of flourishing, perceived quality of life with a specific health condition.

Mine is a proposal for a better conception of validation. According to it validation will be better off when the philosophical heart of this enterprise is out in the open, articulated, and systematic. In practice my proposal calls for inclusion of philosophers in the process. But I do not have a full story about how to organise this process at the level of scientific institutions. Currently several initiatives are underway to develop checklists, hierarchies, expert panels, common conventions.[13] These initiatives include not just oversight of validity but also of other practical features of measures—reliability, responsiveness, ease of use. Minimally I advocate that these initiatives include something like a conception of validity I proposed.

This conception can also be inserted into publication and refereeing conventions. I have no illusions about how hard it will be for scientists to engage with the normativity of their categories. But a concrete first step would be for research papers to have a section titled 'Normative Validity', which would address whether or not the measure arrived at by the standard methods (those based on evidential subjectivism and correlation-mongering) does not have obvious problems from a normative point of view. Does a measure of happiness, for instance, fail to mention freedom from stress as the PANAS does? If so, that is a point against it. It is now commonplace to require that philosophers relying on empirical assumptions must engage with the relevant scientific literatures. Scientists engaged in validation of well-being measures—one of the most philosophical of scientific tasks—also should acquaint themselves with the relevant philosophical work. And where this philosophical work is missing—for example, due to lack of relevant mid-level theories—this absence should be flagged and urgently addressed.

13. McClimans and Browne (2011) review three such initiatives. Organisation for Economic Co-Operation and Development (2013) is a pioneering effort on regulating measures of subjective well-being.

In calling for this extra constraint on measure validation, I have no ambitions of being a philosopher queen. I make no calls to scrap the existing practices entirely. But adding a normative dimension to the process of measure validation presents no danger of philosophy monarchy. There is a lot of philosophy that the science of well-being can safely ignore, but not this sort—not if we have already agreed that the categories of the science of well-being are thick with values and that the science of well-being is rife with policy ambitions.

AFTERWORD

A short summary is in order. I set out to solve the problem of value-aptness, that is to say how the science of well-being can be convincingly about well-being. The solution would have been more straightforward if well-being was a single concept and a single thing that either did or did not corresponded to the eponymous constructs in science. But I found that conceptions of well-being vary across spheres of life and contexts of knowledge production. I sought to understand why the talk of well-being does that and whether it is a good thing. My hypothesis is that the content of the concept of well-being, at least partly, varies with context. Is there anything that unifies all the diverse meanings? I am happy to allow for a minimal core meaning—well-being is a summary value of goods important to the agent for reasons other than moral, aesthetic and political. This is not much; it is perhaps a historical and cultural accident that a single word exists to mark them all. It is no surprise then that to know what states realise this diversity of concepts we shall likely need more than one substantive theory of well-being. A unification, so important to philosophers, seems unlikely and unnecessary in the sciences. The theories that the science really needs are neither unified nor intricate but rather those that tell us the basic ingredients of well-being in different situations for different kinds of creatures—mid-level theories. When we try to build one, as I have for the case of children, we see that high theories are incredibly useful as raw material but also insufficient. Mid-level theories could really improve the science of well-being, they could give it value-aptness, an assurance that the target of research is conceptualised in the right way. But there is an obstacle to my proposal: a reluctance to let values into science. To remove this obstacle I have tried to show that value-laden categories present no threat to objectivity of science, nor does this objectivity require heavy-duty metaphysics: rather than discovering what well-being truly is, we need only subject the controversial aspects of well-being definitions to duly inclusive deliberation. Measurement is also a realistic ideal. Against the claim that well-being is unmeasurable I argued that it could be, provided that we predicate well-being of kinds of people in specific circumstances. On this picture the science of well-being is unlikely to speak directly about all-things-considered individual well-being but can nevertheless supply knowledge about well-being in its various senses. While I am pessimistic about ambitious measures of overall individual well-being, I am optimistic about measures of contextual well-being and measures of well-being components, such as happiness, mental health, meaningful work, and so on. This goal requires an explicit importation of mid-level theories of well-being into the process of measure development and validation. The current status quo on which validation is a technocratic exercise that hides or avoids values is inadequate. Nevertheless my picture is broadly optimistic. The science of well-being can be value-laden, objective, and empirical at once.

I hope to have accomplished that much. Let me now look forward to the unfinished business.

Chief among the unanswered questions is whether the science of well-being is a morally and politically justified pursuit. This field is riding a wave of popularity and excitement for reasons that are not altogether uplifting, as Will Davies (2015) explores in his book *The Happiness Industry: How the Government and Big Business Sold Us Well-Being.* On the business side, the science of well-being is propped up by a massive and lucrative industry that seeks to manage consumption and work in an increasingly automated and data-heavy capitalist economy—workers must be happy at their jobs no matter what; consumers must be happy *and* unhappy just enough to keep consuming. As a popular saying in the management literature goes: 'what gets measured, gets managed'.[1] On the government side, positive psych-ology coupled with technologies of surveillance can be seen in the context of 'personalisation of the public interest'. As such it offers a tempting path for dismissing social problems by making them a mat-ter of *individuals* mismanaging their happiness rather than *communities* mismanaging their politics (Davies, 2011, 2015).

I am less pessimistic than Davies. It should be possible to practice the science of well-being in a responsible and morally aware way. It should be possible to divorce its important findings from the frequently unsavoury roots and fame of this research. But that would call for a broader conception of value-aptness than I have represented in this book. I took value-aptness to consist in the choice of the right construct of well-being for the context, and I defined this rightness in terms of the corresponding mid-level theory. This sense of value-aptness remains necessary, but it may prove to be too narrow of a basis on which to judge a given scientific project. As well as paying attention to values underlying the choice of the construct scientists should watch what they are doing in a broader way, because the science of well-being can be globally sinister even when it is locally innocent. Whether scientists have a responsibility for anticipating the misuses of their discoveries about well-being and how they should react to these potentialities are, of course, difficult questions that admit no single answer. These ethics have yet to be worked out.

Similarly yet to be worked out are the epistemic requirements for the use of this knowledge. It is one thing to have value apt findings and another to put them to use in a way that will work. In the Introduction I urged that we need to practice science and philosophy in a joined up manner. But in fact, as Nancy Cartwright (2006) has recently urged, there are three ingredients, not two: science, philosophy, and use. Whether a given policy/therapy/intervention on well-being does the job depends on more than just the confirmation of this knowledge given the standard academic tests. Ultimately it is the details of the implemen-tation that matter. The same holds for measures of well-being—they are

1. The origins of the quote are unclear going back, Conrad Heilmann tells me, probably to Peter Drucker's 'management by objectives' approach.

not simply valid even when informed by the right mid-level theory—they need to be valid in the context in which they will be used. This territory is entirely unexplored in today's academia. There is an optimistic welfarism permeating the writings by scientists—if wealth and well-being sometimes diverge, how can anyone deny that well-being is a permissible goal of policy, at least in addition to the conventional economic ones? After all, once we can measure well-being and know its causes, and once we show that economic indicators do not always capture it, what else is needed to justify applying this knowledge?

Well, a lot actually. Exactly how well-being is used for planning and evaluation matters a great deal. Benchmarking in healthcare and education provides a wealth of depressing examples for how the pursuit of targets in the guise of accountability introduces perverse incentives and destroys trust and quality for all involved (Bevan & Hood, 2006). Not much is known about the ways in which well-being targets could work out in practice, caution is warranted more than ever, and the input of political scientists is crucial.

These anxieties notwithstanding, misery and flourishing are knowable and too important for science to ignore. So my enthusiasm for this field is undampened, and neither I hope is that of my readers.

APPENDIX A

An Introduction to Philosophy of Well-Being

In a short but famous Appendix I to his *Reasons and Persons*, Derek Parfit (1984, pp. 493–502) discussed 'theories about self-interest' titling the section 'What Makes Someone's Life Go Best'. This focus—on the noninstrumental value to the person, also known as prudential value—characterises the distinct preoccupation of analytic philosophers who write on well-being. Parfit distinguished between mental state, desire fulfillment, and 'objective-list' theories of well-being. This way of carving up the options, what I have called the Big Three, remains useful notwithstanding other more precise classifications emerging (Haybron, 2008, Chapter 2; Woodard, 2013).

Let us be clear at the outset what sort of debate philosophers are engaged in. By and large, the debate is not about what sort of life to pursue and what choices to make in order to be well. It is not, or at least rarely, a *how-to* debate. So people in crisis wishing to reform their lives for better would be ill advised to look for help here. (This is the self-proclaimed goal of positive psychology, life coaching, and psychotherapy.) Indeed, in the vast majority of cases philosophers do not seek to judge whether a particular life or a particular choice is good or bad for an individual. This is taken to be an applied question that can only be answered once a theory of prudential value is specified. This specification at a fairly abstract level is taken to be the business of philosophers. The application of theories—the business of others.

This consensus is only just beginning to be challenged. Valerie Tiberius (2008), for instance, distinguishes between 'target' discussions of well-being (the classic focus I identified earlier) and 'process' discussions. The latter would be a properly how-to story—what sort of values and attitudes should regulate my life?—of which Tiberius offers one. Michael Bishop (2014) also tries to upend the philosophical status quo by identifying well-being with a causal network that involve elements of all three components identified by the Big Three—happiness, success in goals, positive functioning. Well-being is the groove where all these elements keep causing each other and none is more essential than the others (Bishop, 2014). Both of these are attempts to reorient philosophy of well-being and I trust they will be consequential. But for purposes of this overview I concentrate on the standard target theories.

There are, at least, two goals for such a theory. It should list the noninstrumentally valuable goods, and it should identify what makes any valuable good from that list valuable. Roger Crisp (2006) calls the first project *enumerative* and the second *explanatory*. This is important because the good we might specify as an essential ingredient of well-being in our enumeration might be prudentially valuable for a reason that is conceptually distinct. It might be good for us to, say, exercise our capacity to love because it fulfils our nature or satisfies desire or is pleasurable. So the answers to enumerative and explanatory questions may be very different.

Though both questions are discussed by philosophers, it is fair to say that the explanatory question takes up more of their time and energy than the enumerative. Most of the action and debate are generated by the questions about the properties of valuable goods that make them valuable in the right way. To answer this question philosophers must inevitably say something about the goods that have this property—which requires delving into the enumerative issues—but they do not need to say all that much. Enjoyment is such a good, as is friendship, but what sort of enjoyment exactly and what sort of friendship is beyond what is currently discussed, at least in the literature on prudential value. So the main disagreement is about why we need what we need. Very roughly, for subjectivists it is because we want these goods, for hedonists it is the way they make us feel, and for objectivists it is the way they suit our nature.

Hedonists take our mental states to hold the key prudential property and only them—not just any mental state, of course, but only experiences with a positive valence. What states exactly? Hedonists take the relevant state to be pleasure, or satisfaction, or enjoyment, which for present purposes are synonymous. A recent example of a hedonist theory of well-being comes from Roger Crisp (2006, p. 622): 'what is good for any individual is the enjoyable experience in her life, what is bad is the suffering in that life, and the life best for an individual is that with the greatest balance of enjoyment over suffering'. Other versions of hedonism have recently been proposed by Feldman (2002), Bradley (2009), Bramble (2016). Explanatory hedonists can accept that things other than enjoyment can be good for us but only in virtue of their enjoyability. Great art, friendship, virtue can all benefit us but only via their causal effect on our experiences.

However, much rides on how we define enjoyment. What makes an experience pleasurable? Is it pleasurable in virtue of how it feels or in virtue of our liking it? Very roughly these positions are respectively *internalist* and *externalist* (Sumner, 1996). They are called so because in the first case pleasures are identified by their internal quality whereas in the second—by an attitude external to the pleasure—our liking it. For original internalists such as Jeremy Bentham, pleasure was a special, unanalysable quality of an experience, the feeling-good quality. For original externalists such as Henry Sidgwick, who could not find any common quality in pleasures as distinct as, say, eating and singing, pleasure was to be identified by the fact that it is desired by an agent. But later internalists thought that it is perfectly conceivable not to desire a pleasure and searched for another internalist solution. Perhaps pleasure is like volume in that it is not a single quality but a single dimension along which sounds vary (Kagan, 1992), or perhaps we should identify pleasure by the neurochemical processes in the brain that appear to underlie enjoyments of very different kinds (Crisp, 2006). Finally, externalists reply with further options for the correct propositional attitude to identify pleasure—liking, desiring, favouring (Feldman, 2002; Heathwood, 2007).

The outcome of this debate is not trivial, for externalists about pleasure have more affinities with desire fulfillment theories than with hedonism. If pleasure is that which we desire, then the hedonist view that pleasure is good for us becomes a version of a desire satisfaction view, that is, the view that it is good for us to satisfy our desires, of which pleasure is (perhaps the only) one. But in that case why focus only on pleasures? We might as well conceive of well-being as having access to *any* desired object. If we also specify that desires must be actually rather than subjectively satisfied, we would be squarely in the territory of desire fulfillment or more generally subjectivist views of well-being.

This view prides itself on not falling victim to the experience machine argument first put forward by Robert Nozick. Take two people with identical experiences. One lives them 'for real', the other by being connected to a machine that simulates his brain in precise ways. Pure hedonists have no option but to bite the bullet and admit that neither life is any more prudentially valuable. The desire fulfillment theorist can claim that the denizen of the experience machine has not *really* had their desires fulfilled; they only think they have.

This view has another advantage. For something to be good for me, this good has to have a special relationship to me, it has to engage me, or resonate with me, or be responsive to my priorities, or some such. This is referred to in various ways as the resonance constraint (Brink, 2008), or the agent sovereignty (Arneson, 1999), or the subject-relativity (Sumner, 1996). Classical hedonism takes pleasure to be good for me without consulting me. The thought experiments of monks or tortured artists with no desire (nor other kinds of relevant attitude) for pleasure and yet a great satisfaction with their lives are meant to bring out the intuition that desire-based theories are superior to hedonism.

According to a basic desire fulfillment view, it is good for us to get what we want—actually get it, not just think we do—and that is the only thing that is good for us. (Or on the explanatory version desire fulfillment is the only property that makes any good prudentially valuable.) But sometimes people want things for themselves that seemingly do them no good whatsoever. Perhaps they are uninformed, or indoctrinated, or perhaps their desires are only for things that have little to do with their lives, like survival of polar bears, or trivial things like another chewing gum. For those cases, there are various bells and whistles. First we can say that it is not *desire* fulfillment but *goals* or *value* fulfillment that matters (Dorsey, 2012; Scanlon, 1998; and many others). Second, we can restrict which desires or goals or beliefs count, any actual ones (Keller, 2009) or perhaps only idealised ones, for instance only those that one would have after a good reflection (Brandt, 1979) or only those that a fully informed agent would want their actual self to have (Railton, 1986). These are known as *idealised* or full information versions. Third, we can restrict the content of these valuing attitudes, for example to concern only life projects (Dorsey, 2009).

We can even try to tweak the theory to exclude John Rawls's famous grass-counter—the person who with full information and sincerity announces that his goal in life is to count blades of grass on a lawn. He does so and claims to be doing perfectly well. But if we want our theory of well-being to exclude the grass-counter we might as well admit to being, at least partly, objectivist, believing that well-being can encompass goods that benefit (or harm) a person no matter what their attitudes, life plans, or tastes are. Perhaps the most famous such theory is Aristotle's perfectionism—the best life for a person is to function at the highest level a normal human could, which involves exercising distinctly human virtues of justice, friendship, contemplation, and so on—a state he called eudaimonia (or, on the explanatory version, prudentially valuable life is such because it fulfils our nature). There are ancient versions of this theory (Annas, 1993) and several modern ones. Stephen Darwall's (2002) neo-Aristotelian proposal is that well-being consists in 'valuing activities', that is, the activities that bring us into contact with objects of worth, where worth is understood independently of anyone's desires. Understanding art, bringing up children, and building a relationship are all examples of valuing activities. Richard Kraut (2007) maintains another Aristotelian view that emphasises development of skills appropriate to the being's nature and stage of growth. There are many options in this tradition. For Haybron (2008, Chapter 9), well-being is living in accordance with one's own emotional nature—a view he calls self-fulfillment, for Neera Badhwar (2014)—in accordance with moral virtue.

There are also objectivists with no specific links to Aristotle who take well-being to be constituted by goods that are prudentially valuable no matter the agent's or anyone else's attitude. These goods are part of a list arranged in order of importance, hence the name objective list theory of well-being (Fletcher, 2013, 2016). The capabilities approach that I discuss in Appendix B is to my knowledge the only operationalisation of an objective list view.

Finally, there are hybrid theories of well-being, joining two or more of these accounts into one. It is hard to distinguish hybrids between hedonism and objective list on the one hand from those objective list theories that list enjoyment as one of the goods (Darwall, 2002; Griffin, 1986; Parfit, 1984). There are also hybrids of subjectivism and hedonism (Hawkins, 2010; Heathwood, 2007). Hybrid theories are not a panacea though; they face distinct challenges of explaining the structure and relationship between the different goods (Sarch, 2012; Woodard, 2015). One way to avoid hybridity is to show that the subjective and objective elements are not separate. This is the conceit of Simon Keller's success theory:

> An individual has a high level of welfare to the extent that she is successful, in a certain sense. To be successful in that sense is to have attitudes that do well according to the standards they constitutively set for themselves. Three species of such attitudes are beliefs, goals, and evaluative attitudes. There may also be a fourth such attitude: the attitude of immediately liking an experience. (Keller, 2004, p. 668)

The proposal is ingenious for trying to accommodate both the agent relativity, by its reference to attitudes, and objectivity, by invoking constitutive standards for these attitudes.

APPENDIX B

Constructs of Well-Being across Sciences

Table I.1 in the Introduction is my classification of constructs of well-being across the social and medical sciences. This appendix serves as an explanation of this table. I skip the row 'Child Sciences' since it has a dedicated chapter. My classification is in terms of the academic disciplines and the theories of well-being that inspire the constructs used in each discipline. For an alternative way of classifying well-being constructs see Gasper (2010).

PSYCHOLOGICAL SCIENCES

Throughout the book I refer to three approaches to well-being in psychology—happiness, life satisfaction, and flourishing. I discuss the traditions and their methods of measurement thoroughly in the main text so will be brief here. All of these constructs are subjective but all in different senses. Happiness is hedonic or affective profile, life satisfaction—a favourable judgement of one's life, and flourishing—a sense of meaning and accomplishment. Although each is sometimes used as a representation of well-being on its own, it is increasingly common for psychologists to study the relations between the three and to theorise about their union (Arthaud-Day, 2005; Kahneman & Riis, 2005; Kim-Prieto, 2005). The term 'subjective well-being' refers sometimes to one of the three and sometimes to their union.

ECONOMIC SCIENCES

Well-being concepts enter into today's economics in several ways:

- In fundamental theory of microeconomics utility is sometimes understood as a quantity denoting well-being and via this route well-being enters theoretical and empirical studies of judgement, choice, and strategic interaction.
- Measures of growth in macroeconomics sometimes refer to well-being, especially in discussions of goals of development. Thus research around Easterlin Paradox yielded estimates of how much subjective well-being does or does not react to income.
- For some economists well-being is the benefit in cost-benefit analysis and related exercises.

Debates on all these issues are animated in part by the contrast between the orthodoxy and the alternatives. Traditionally economics operated with a preference satisfaction view, which is a version of philosophical subjectivism, but does not restrict preferences in any way. What we want is what is good for us. Welfare economics is a theoretical system based on this simple (to many philosophers, dangerously simple) view of well-being. Moreover, this view is standardly supplemented with a definition of preferences as revealed choices. What we want is, roughly, what we choose when given an opportunity. Whether the choices are horrible, manipulative, or whether we ourselves are deceived, weak-willed, or irrational does not matter on this view. This is the actual, rather than idealised, preference satisfaction. This package of ideas, as well as underpinning welfare economics, is also the prime input into evaluation of social states by means of cost-benefit analysis. The value of alternative courses of actions (to pursue or not to pursue a given policy) is their present monetary value as revealed by market prices of similar states, willingness to pay, or stated preferences. For instance, the UK Treasury's Green Book mandates this procedure for all policy evaluation (HM Treasury, 2003). Exactly how to measure benefit, living standards, and poverty using these methods is a research program on its own. National account such as gross domestic product often give strikingly different answers than surveys of household consumption, even though both are inspired by the preference satisfaction tradition (Deaton, 2016).

This tradition is alive and well. Writing for *The New York Times*, a Harvard economist Edward Glaeser calls this view 'the moral heart of economics'. He elaborates:

> improvements in welfare occur when there are improvements in utility, and those occur only when an individual gets an option that wasn't

previously available. We typically prove that someone's welfare has increased when the person has an increased set of choices. When we make that assumption (which is hotly contested by some people, especially psychologists), we essentially assume that the fundamental objective of public policy is to increase freedom of choice. (Glaeser, 2011)

As Glaeser acknowledges, this project is under pressure from many sources, especially its assumption that people have stable and consistent preferences and act rationally so as to get the most of what they prefer. The main source is the empirical research into various biases and irrationalities that afflict choices of actual human beings. Psychologists and behavioural economists have been studying these biases since the 1970s. This research has undermined the assumption of stable preferences and choices that maximise them and demonstrated the distinction between experienced utility and decision utility (Kahneman, 1999 among others). As a result of these studies, even mainstream economists now recognise that actual choices do not always reveal what is truly good for people. But what other conception of well-being should economists turn to?

There are roughly three reactions to this challenge (in addition to ignoring it). One is to supplement the standard economic framework with data on subjective well-being as we have seen psychologists to advocate. Fujiwara and Campbell (2011) writing for the UK Treasury explain how to use life satisfaction data in standard cost-benefit analysis. The Organisation for Economic Co-operation and Development's (OECD, 2013) guidelines also advocate this.

The second reaction is to clean up the preferences and the conditions under which they are revealed. Only some preferences and only those revealed when people make thoughtful and important decisions are indicative of their well-being (Hausman & McPherson, 2009). Some such selection is already part of normal cost-benefit analysis (Adler & Posner, 2006), and some takes creative new methods (Benjamin et al., 2014). Importantly, both normative and empirical assumptions go into specifying the circumstances in which choices are authoritative of 'clean' preferences. Choices should not be too far in the future; they should be about important rather than trivial things, with proper reflection and ideally with prior practice (Benjamin et al., 2014; Beshears et al., 2008).

The third reaction is to move away from preferences altogether. Not all economists are wedded to a preference satisfaction view. Development economics, for instance, has its own robust tradition of theorising about well-being along entirely different lines.

The capabilities approach is associated with economist-philosopher Amartya Sen and philosopher Martha Nussbaum as a framework for measuring justice, development, and progress (Nussbaum, 2000; Nussbaum & Sen, 1993; among other sources). The idea is that humans need the freedom to pursue distinct capabilities, their 'beings and doings' (Robeyns, 2011), which may include health, education, political rights, social relationships, emotional life, creativity, and so on. Human well-being is here understood as an ability to achieve valuable states, rather than the traditional economic utility. Capabilities are different from utility in that their value cannot be measured on a single scale and as a result they do not admit trade-offs (sacrificing, say, political rights for access to healthcare cannot make a person better off). They also make room for the fact that different people might need different amounts of goods or services to fulfil a capability. Capabilities are typically defined using a theory of objective human needs (e.g., an Aristotelian theory) rather than by consulting people's preferences, let alone the ones revealed by choices. The capabilities approach inspired the United Nations Development Project's Human Development Index, now more than 20 years old and still serving as the measure of progress and development (Anand & Sen, 1994).

Even economists who do not subscribe to the capabilities approach advocate an objective understanding of well-being for development contexts. Partha Dasgupta (2001) proposes the notion of *aggregate quality of life*. It is aggregate in two senses: first, it represents the state of many people, and, second, their quality of life is constituted by several elements. Dasgupta (2001, p. 54) writes: 'a minimal set of indices for spanning a reasonable conception of current well-being in a poor country includes private consumption per head, life expectancy at birth, literacy, and civil and political liberties'. Private consumption is food, shelter, clothing, and basic legal aid. Life expectancy at birth is the best indicator of health, while literacy of basic primary education. Civil and political rights allow people to function independently of the state and their communities. Each of these is necessary. They cannot be reduced to some one item or replaced by a monetary value, for they may be undervalued by the market.

However, current quality of life is not the only thing we mean when we ask 'How well is a country doing?' Sometimes we also mean to inquire about what Dasgupta calls a country's *social well-being*. This concept encompasses, along with the current quality of life, the *sustainability* of this current lifestyle—how well does a country balance the needs of its current population with the needs of its future generations? A high quality of life at a time may conceal the fact that a community is consuming its resources without an adequate provision for the future. For Dasgupta, social well-being is a pattern of consumption that strikes the best balance between current and future quality of life. Its cause and the best measure is a country's *wealth*, best represented by an index comprising the social value of a country's natural resources, manufactured capital, its human and social capital (which includes public knowledge, institutions, etc.), minus this country's liabilities.

The idea of sustainability enters also new proposals for national well-being. Here we see a genuine mixture of different traditions (economics, psychology, other social sciences).

NATIONAL WELL-BEING

That traditional economic measures are inadequate for capturing national well-being was eloquently argued by Bobby Kennedy in his oft-cited 1968 address at the University of Kansas, Lawrence:

> Our gross national product . . . counts air pollution and cigarette advertising, and ambulances to clear our highways of carnage. It counts special locks for our doors and the jails for those who break them. It counts the destruction of our redwoods and the loss of our natural wonder in chaotic sprawl. Yet the gross national product does not allow for the health of our children, the quality of their education, or the joy of their play. It does not include the beauty of our poetry or the strength of our marriages; the intelligence of our public debate or the integrity of our public officials. It measures neither our wit nor our courage; neither our wisdom nor our learning; neither our compassion nor our devotion to our country; it measures everything, in short, except that which makes life worthwhile.

What measure would capture that which makes life worthwhile at the level of a nation? That it should include more than the traditional economic indicators is slowly becoming the mainstream view. In 2009 three eminent economists Joseph Stiglitz, Amartya Sen, and Jean Paul Fitoussi produced a report commissioned by the then French President Nicolas Sarkozi outlining a multi-dimensional measure of national well-being that includes even subjective well-being indicators (Stiglitz et al., 2009). As we have seen in the Introduction and Chapter 4, all three traditions in psychology, plus the two in economics, have sought to contribute their elements to the overall metric. In case of United Kingdom they have succeeded; the Office of National Statistics' measure of national well-being is a mongrel with room for all traditions.

Two requirements seem to be crucial to a measure of national well-being. First, such a measure needs to capture the values and priorities of the people whose well-being it is supposed to represent. Haybron and Tiberius (2015) coined the term *pragmatic subjectivism* precisely for this purpose. They argue that even if one adopts an objectivist theory of well-being, when it comes to well-being policy at a governmental level one should adopt a kind of subjectivism—not an actual preference satisfaction view but a more sophisticated subjectivism: one that differentiates between stated or revealed preference and deeply held values and prioritises the latter. Because policy contexts present special dangers of paternalism and oppression, governments defer on the nature of well-being to the individuals they represent. (None of this implies that governments should stay out of promoting well-being of its citizens). Second, a measure of national well-being needs to represent a certain level of consensus, not a mere sum of individual well-beings. This is what the Office of National Statistics (ONS) sought in its 2011 public consultations. Other such initiatives are Germany's 2015 'Gut leben in Deutchland' and France's 'Nouveaux indicateurs de richesse'. Together these two requirements appear to underlie the rationale behind these approaches and other multi-indicator proposals such as the Prosperity Index of the Legatum Institute (Legatum, 2015) and the OECD's Better Life Index.

MEDICAL SCIENCES

Is well-being a goal of medicine and healthcare? There are strikingly diverse answers to this question (Groll, 2015; Hausman, 2015). But in today's medical sciences the stand-in for well-being is *health-related quality of life* (HRQL), and I aim only to summarise how it is measured. There are roughly two institutional traditions.

For health economists, HRQL is an input into the calculation of Quality Adjusted Life Years (QALY), which are in turn used in some countries to judge the relative efficiency of different ways of allocating healthcare (e.g., National Institute of Health and Clinical Excellence, 2013). The official goal is to apportion limited resources in a way that best answers to the wishes of the taxpayers and justice, where cost per QALY is an indicator of efficiency (Nord, 1999). In this tradition HRQL is gauged by questionnaires such as EQ5D and it reflects respondents' ranking of many different health states according to their preferences. This ranking is inferred from the way in which people on average trade off different health states (as indicated by their stated preferences between, say, being mildly depressed for a year and having a broken hip). Today QALYs are losing their popularity somewhat under pressure from two sources: the weakened confidence in the existence of well-behaved utility functions and increasing evidence that these preferences are informed by ignorance, prejudices and fears about the nature of illness and disability (Carel, 2016; Hausman, 2015). Subjective well-being measures which gauge how patients with different conditions feel rather than health states they would prefer are now entering this tradition too (Dolan, 2000).

The second approach comes from the field of health services management (and to a lesser extent health technology assessment). Driven more by medical administrators and researchers, this tradition measures HRQL with questionnaires that reveal patients' own evaluation of their state conceived as a combination of subjective satisfaction and self-reported functioning, adjusted specifically by age, circumstances, and the specific illness. This information features in patient-reported outcomes, and it is used for evaluating the effectiveness of interventions, whether in clinical trials or in clinical practice (Fayers & Machin, 2013; Guyatt et al., 1993). The hope is to represent the patients' perspective on their condition and the treatment they receive. So researchers use questionnaires about the health status in the tradition of psychology as described in Chapters 5 and 6.

Some of these questionnaires are generic and others are disease specific. Generic questionnaires access the respondent's health status as a whole, taking into account their functioning along all the main dimensions of daily life. Instruments such as the World Health Organization Quality of Life questionnaire, the Nottingham Health Profile, and the Sickness Impact Profile provide a general picture of the subject's health, both subjective and objective, including pain, symptoms, and psychosocial and environmental stressors (Bergner et al., 1981; Hunt et al., 1985). Nongeneric measures are developed for people with a specific illness or in specific circumstances. QUALEFFO and Caregiver Strain Index, the two examples I discuss in Chapter 2, were designed respectively for people with vertebral fractures and osteoporosis and caretakers of spouses with heavy chronic illness (Gerritsen & Van Der Ende, 1994; Lips et al., 1997, 1999). Freedom from the caregiver strain is then combined with life satisfaction to make a special measure of well-being for caregivers (Stull, 1994; Takai et al., 2009; Visser-Meily et al., 2005). Nongeneric measures are prized for their specificity: they 'may be more valid, in the sense that they measure quality of life more accurately in that particular disease than generic instruments' (Lips et al., 1999, p. 151). When there is a change in the patient's state, these measures supposedly pick it up more readily than the generic instruments, an important though disputed virtue in clinical studies (Dowie, 2002; Fitzpatrick et al., 1992; Fletcher et al., 1992).

WORKS CITED

Adler, A., & Seligman, M. E. (2016). Using wellbeing for public policy: Theory, measurement, and recommendations. *International Journal of Wellbeing, 6*(1), 1–35.

Adler, M. D., & Posner, E. A. (2006). *New foundations of cost-benefit analysis.* Cambridge, MA: Harvard University Press.

Alexandrova A., & Haybron, D. (2016). Is construct validity valid? *Philosophy of Science, 83*(5), 1098–1109.

Anand, S., & Sen, A. (1994). *Human Development Index: Methodology and measurement* (No. HDOCPA-1994-02). New York: Human Development Report Office, United Nations Development Programme.

Anda, R. F., Whitfield, M. C. L., Felitti, V. J., Chapman, M. D., Edwards, V. J., Dube, P. D. S. R., & Williamson, D. F. (2002). Adverse childhood experiences, alcoholic parents, and later risk of alcoholism and depression. *Psychiatric Services, 53*, 1001–1009.

Anderson, E. (2004). Uses of value judgments in science: A general argument, with lessons from a case study of feminist research on divorce. *Hypatia, 19*(1), 1–24.

Anderson, E. (2006). How not to criticize feminist epistemology: A review of Scrutinizing feminist epistemology. Retrieved from http://www-personal.umich.edu/{~{}}eandersn/hownotreview.Html

Anderson, E. (2014). *Social movements, experiments in living, and moral progress: Case studies from Britain's abolition of slavery.* Lindley Lecture Series. Lawrence: University of Kansas.

Angner, E. (2009). Subjective measures of well-being: Philosophical perspectives. In H. Kincaid & D. Ross (Eds.), *The Oxford handbook of philosophy of economics* (pp. 560–579). Oxford: Oxford University Press.

Angner, E. (2011a). Current trends in welfare measurement. In J. B. Davis & D. W. Hands (Eds.), *The Elgar companion to recent economic methodology* (pp. 121–54). Northampton, MA: Edward Elgar. doi:10.4337/9780857938077.00012

Angner, E. (2011b). The evolution of Eupathics: The historical roots of subjective measures of wellbeing. *International Journal of Wellbeing, 1*(1), 4–41.

Angner, E. (2013). Is it possible to measure happiness? *European Journal for Philosophy of Science, 3*(2), 221–240.

Angner, E. (2015). Well-Being and economics. In G. Fletcher (Ed.), *The Routledge handbook of the philosophy of well-being* (pp. 492–203). London: Routledge.

Annas, J. (1993). *The Morality of Happiness*. New York: Oxford University Press.

Archard, D. W. (2011). Children's rights. In Edward N. Zalta (Ed.), *Stanford encyclopedia of philosophy*. Retrieved from http://plato.stanford.edu/archives/sum2011/entries/rights-children/

Arneson, R. (1999). Human flourishing versus desire satisfaction. *Social Philosophy and Policy, 16*(1), 113–142.

Arthaud-Day, M. L., Rode, J. C., Mooney, C. H., & Near, J. P. (2005). The subjective well-being construct: A test of its convergent, discriminant, and factorial validity. *Social Indicators Research, 74*, 445–476.

Axford, N. (2009). Child well-being through different lenses: Why concept matters. *Child and Family Social Work, 14*, 372–383.

Badhwar, N. K. (2014). *Well-being: Happiness in a worthwhile life*. Oxford: Oxford University Press.

Beddington, J., Cooper, C. L., Field, J., Goswami, U., Huppert, F. A., Jenkins, R. ... Thomas, S. M. (2008). The mental wealth of nations. *Nature, 455*(7216), 1057–1060.

Ben-Arieh, A. (2006). Is the study of the "state of our children" changing? Revisiting after 5 years. *Children and Youth Services Review, 28*(7),799–811.

Ben-Arieh, A. (2010). Developing indicators for child well-being in a changing context. In C. McAuley & W. Rose (Eds.), *Child well-being: Understanding children's lives* (pp. 129–142). London: Jessica Kingsley.

Benjamin, D. J., Heffetz, O., Kimball, M. S., & Szembrot, N. (2014). Beyond happiness and satisfaction: Toward well-being indices based on stated preference. *American Economic Review, 104*(9), 2698–2735.

Bergner, M., Bobbitt, R. A, Carter, W. B., & Gilson, B. S. (1981). The sickness impact profile: Development and final revision of a health status measure. *Medical Care, 19*, 787–805.

Beshears, J., Choi, J. J., Laibson, D., & Madrian, B. C. (2008). How are preferences revealed? *Journal of Public Economics, 92*(8), 1787–1794.

Bevan, G., & Hood, C. (2006). What's measured is what matters: Targets and gaming in the English public health care system. *Public Administration, 84*(3), 517–538.

Bishop, M. (2015). *The good life: Unifying the philosophy and psychology of well-being*. Oxford: Oxford University Press.

WORKS CITED

Blackburn, S. (2013). Disentangling. In S. Kirchin (Ed.), *Thick concepts* (pp. 267–283). Oxford: Oxford University Press.

Borsboom, D. (2005). *Measuring the mind: Conceptual issues in contemporary psychometrics.* Cambridge, UK: Cambridge University Press.

Boyd, R. (1991). Realism, anti-foundationalism and the enthusiasm for natural kinds. *Philosophical Studies, 61,* 127–148.

Bradley, B. (2009). *Well-being and death.* Oxford: Oxford University Press.

Bradley, R. H., & Corwyn, R. F. (2000). Moderating effect of perceived amount of family conflict on the relation between home environmental processes and the well-being of adolescents. *Journal of Family Psychology, 14*(3), 349–364.

Bradley, R. H., & Corwyn, R. F. (2002). Socioeconomic status and child development. *Annual Review of Psychology, 53,* 371–399.

Bradshaw, J., Hoelscher, P., & Richardson, D. (2007). An index of child well-being in the European Union. *Social Indicators Research, 80*(1), 133–177.

Bramble, B. (2016). A new defense of hedonism about well-being. *Ergo, 3*(4), 85–112.

Brandt, R. (1979). *A theory of the good and the right.* Oxford: Clarendon Press.

Brennan, S. (2002). Children's choices or children's interests: Which do their rights protect? In C. Macleod & D. Archard (Eds.), *The moral and political status of children: New essays* (pp. 53–69). Oxford: Oxford University Press.

Brennan, S. (2014). The goods of childhood, children's rights, and the role of parents as advocates and interpreters. In F. Baylis & C. McLeod (Eds.), *Family-making: Contemporary ethical challenges* (pp. 29–48). Oxford: Oxford University Press.

Brighouse, H. (2003). How should children be heard? *Arizona Law Review, 45*(3), 691–711.

Brighouse, H., & Swift, A. (2014). *Family values.* Princeton, NJ: Princeton University Press.

Brink, D. O. (2008). The significance of desire. *Oxford Studies in Metaethics, 3,* 5–45.

Brink, D. O. (2013). *Mill's progressive principles.* Oxford: Oxford University Press.

Brogaard, B. (2008). Moral contextualism and moral relativism. *The Philosophical Quarterly, 58*(232), 385–409.

Brooks-Gunn, J., & Duncan, G. J. (1997). The effects of poverty on children. *The Future of Children, 7*(2), 55–71.

Brooks-Gunn, J., Duncan, G. J., & Aber, J. L. (1997). *Neighborhood poverty.* New York: Russell Sage Foundation.

Broome, J. (2004). *Weighing lives.* Oxford: Oxford University Press.

Brown, M. J. (2013). Values in science beyond underdetermination and inductive risk. *Philosophy of Science, 80*(5), 829–839.

Camfield, L., Crivello, G., & Woodhead, M. (2009). Wellbeing research in developing countries: Reviewing the role of qualitative methods. *Social Indicators Research, 90*(1), 5–31.

Campbell, S. M. (2013). An analysis of prudential value. *Utilitas, 25*, 334–354.

Campbell, S. M. (2015). The concept of well-being. In Guy Fletcher (Ed.), *Routledge handbook of the philosophy of well-being* (pp. 402–415). London: Routledge.

Caplin, A., & Schotter, A. (Eds.). (2008). *The foundations of positive and normative economics: A handbook*. New York: Oxford University Press.

Cappelen, H., & Lepore, E. (2005). *Insensitive semantics: A defense of semantic minimalism and speech act pluralism*. Malden, MA: Wiley-Blackwell.

Cappelen, H., & Lepore, E. (2007). The myth of unarticulated constituents. In M. O'Rourke & C. Washington (Eds.), *Essays in honor of John Perry* (pp. 199–214). Cambridge, MA: MIT Press.

Carel, H. (2016). *The phenomenology of illness*. Oxford: Oxford University Press.

Cartwright, N. (1989). *Nature's capacities and their measurement*. Oxford: Oxford University Press.

Cartwright, N. (1999). *The dappled world: A study of the boundaries of science*. Cambridge, UK: Cambridge University Press.

Cartwright, N. (2006). Well-ordered science: Evidence for use. *Philosophy of Science, 73*(5), 981–990.

Cartwright, N., & Bradburn, N. (2011). A theory of measurement. In R. M. Li (Ed.), *The importance of common metrics for advancing social science theory and research: Proceedings of the National Research Council Committee on Common Metrics* (pp. 53–70). Washington, DC: National Academies Press.

Cartwright, N., Shomar, T., & Suárez, M. (1995). The toolbox of science: Tools for the building of models with a superconductivity example. *Poznan Studies in the Philosophy of the Sciences and the Humanities, 44*, 137–149.

Casas, F. (2011). Subjective social indicators and child and adolescent well-being. *Child Indicators Research, 4*(4), 555–575.

Chang, H. (2004). *Inventing temperature: Measurement and scientific progress*. Oxford: Oxford University Press.

Chapman, D. P., Whitfield, C. L., Felitti, V. J., Dube, S. R., Edwards, V. J., & Anda, R. F. (2004). Adverse childhood experiences and the risk of depressive disorders in adulthood. *Journal of Affective Disorders, 82*(2), 217–225.

Chase-Lansdale, P. L., Moffitt, R. A., Lohman, B. J., Cherlin, A. J., Coley, R. L., Pittman, L. D., . . . Votruba-Drzal, E. (2003). Mothers' transitions from welfare to work and the well-being of preschoolers and adolescents. *Science, 299*(5612), 1548–1552.

Children's Society. (2013). The Good Childhood Report 2013. Retrieved from http://www.childrenssociety.org.uk/good-childhood-report-2013-online/index.html

Chilvers, J. (2008). Deliberating competence: Theoretical and practitioner perspectives on effective participatory appraisal practice. *Science, Technology & Human Values, 33*(2), 155–185.

Clark, A. E., Layard, R., & Senik, C. (2012). The causes of happiness and misery. In J. Helliwell, R. Layard, & J. Sachs (Eds.), *World happiness report* (pp. 59–89). New York: The Earth Institute, Columbia University.

Crisp, R. (2006). Hedonism reconsidered. *Philosophy and Phenomenological Research, 73*(3), 619–645.

Crisp, R. (2013). Well-being. In Edward N. Zalta (Ed.), *Stanford encyclopedia of philosophy* (Summer 2013 ed.). Retrieved from http://plato.stanford.edu/archives/sum2013/entries/well-being/

Cronbach, L. J., & Meehl, P. E. (1955). Construct validity in psychological tests. *Psychological Bulletin, 52*(4), 281–302.

Darwall, S. (2002). *Welfare and rational care.* Princeton, NJ: Princeton University Press.

Dasgupta, P. (2001). *Human well-being and the natural environment.* Oxford: Oxford University Press.

Daston, L. (1992). Objectivity and the escape from perspective. *Social Studies of Science, 22*(4), 597–618.

Davies, W. (2011). National well-being: The "personalization of the public interest"? Retrieved from https://www.opendemocracy.net/ourkingdom/william-davies/national-wellbeing-%E2%80%98personalisation-of-public-interest%E2%80%99

Davies, W. (2015). *The happiness industry: How the government and big business sold us well-being.* London: Verso.

Deaton, A. (2016). Measuring and understanding behavior, welfare, and poverty. *American Economic Review, 106*(6): 1221–43. doi:10.1257/aer.106.6.1221

Deaton, A., & Stone, A. A. (2016). Understanding context effects for a measure of life evaluation: How responses matter. *Oxford Economic Papers, 68*(4), 861–870.

De Vet, H. C., Terwee, C. B., Mokkink, L. B., & Knol, D. L. (2011). *Measurement in medicine: A practical guide.* Cambridge, UK: Cambridge University Press.

Deci, E. L., & Ryan, R. M. (2008). Hedonia, eudaimonia, and well-being: an introduction. *Journal of Happiness Studies, 9*(1), 1–11.

DeRose, K. (2000). Now you know it, now you don't. In R. Cobb-Stevens (Ed.), *Proceedings of the Twentieth World Congress of Philosophy,* Vol. V: *Epistemology* (pp. 91–106). Bowling Green, OH: Philosophy Documentation Center, Bowling Green State University.

Diener, E. (2012). New findings and future directions for subjective well-being research. *American Psychologist, 67*(8), 590–597.

Diener, E., & Emmons, R. A. (1985). The independence of positive and negative affect. *Journal of Personality and Social Psychology, 47*(5), 1105–1117.

Diener, E., Emmons, R. A., Larsen, R. J., & Griffin, S. (1985). The Satisfaction with Life Scale. *Journal of Personality Assessment, 49,* 71–75.

Diener, E., Heintzelman, S. J., Kushlev, K., Tay, L., Wirtz, D., Lutes, L. D., & Oishi, S. (2016) Findings all psychologists should know from the new science on subjective well-being. *Canadian Psychology.* http://dx.doi.org/10.1037/cap0000063

Diener, E., Lucas, R., Schimmack, U., & Helliwell, J. (2008). *Well-being for public policy.* New York: Oxford University Press.

Diener, E., Ng, W., Harter, J., & Arora, R. (2010). Wealth and happiness across the world: Material prosperity predicts life evaluation, whereas psychosocial prosperity predicts positive feeling. *Journal of Personality and Social Psychology, 99*(1), 52–61.

Diener, E., & Seligman, M. E. P. (2004). Beyond money: Toward an economy of well-being. *Psychological Science in the Public Interest, 5*(1), 1–31.

Diener, E., & Suh, E. M. (Eds.). (2000). *Culture and subjective well-being.* Cambridge, MA: MIT Press.

Diener, E., Wirtz, D., Tov, W., Kim-Prieto, C., Choi, D. W., Oishi, S., & Biswas-Diener, R. (2010). New well-being measures: Short scales to assess flourishing and positive and negative feelings. *Social Indicators Research, 97*(2), 143–156.

Dolan, P. (2000). The measurement of health-related quality of life for use in resource allocation decisions in health care. *Handbook of Health Economics, 1*, 1723–1760.

Dolan, P., & Peasgood, T. (2008). Measuring well-being for public policy: Preferences or experiences? *Journal of Legal Studies, 37*, 5–31.

Dolan, P., & White, M. (2007). How can measures of subjective well-being be used to inform public policy? *Perspectives on Psychological Science, 2*(1), 71–85.

Dong, M., Dube, S. R., Felitti, V. J., Giles, W. H., & Anda, R. F. (2003). Adverse childhood experiences and self-reported liver disease: New insights into the causal pathway. *Archives of Internal Medicine, 163*(16), 1949–1956.

Dong, M., Giles, W. H., Felitti, V. J., Dube, S. R., Williams, J. E., Chapman, D. P. & Anda, R. F. (2004). Insights into causal pathways for ischemic heart disease: Adverse childhood experiences study. *Circulation, 110*(13), 1761–1766.

Doris, J. M., Knobe, J., & Woolfolk, R. (2008). Variantism about responsibility. *Philosophical Perspectives: Philosophy of Mind, 21*, 183–214.

Dorsett, R., & Oswald A. J. (2014). Human well-being and in-work benefits: a randomized controlled trial. IZA Discussion Paper No. 7943. Retrieved from http://ssrn.com/abstract=2396438

Dorsey, D. (2009). Headaches, lives and value. *Utilitas, 21*(1), 36–58.

Dorsey, D. (2012). Subjectivism without desire. *Philosophical Review, 121*(3), 407–442.

Douglas, H. (2004). The irreducible complexity of objectivity. *Synthese, 138*(3), 453–473.

Douglas, H. (2005). Inserting the public into science. In S. Maasen & P. Weingart (Eds.), *Democratization of expertise? Exploring novel forms of scientific advice in political decision-making, sociology of the sciences,* Vol. 24 (pp. 153–169). Dordrecht: Springer.

Douglas, H. (2009). *Science, policy, and the value-free ideal.* Pittsburgh: University of Pittsburgh Press.

Douglas, H. (2011). Facts, values, and objectivity. In I. C. Jarvie& J. Zamora Bonilla (Eds.), *The SAGE Handbook of Philosophy of Social Science* (pp. 513–529). London: SAGE.

Dowie, J. (2002). Decision validity should determine whether a generic or condition-specific HRQOL measure is used in health care decisions. *Health Economics, 11*(1), 1–8. doi:10.1002/hec.667

Dube, S. R., Anda, R. F., Felitti, V. J., Chapman, D. P., Williamson, D. F., & Giles, W. H. (2001). Childhood abuse, household dysfunction, and the risk of attempted suicide throughout the life span: Findings from the Adverse Childhood Experiences Study. *Journal of the American Medical Association, 286*(24), 3089–3096.

Dube, S. R., Anda, R. F., Felitti, V. J., Edwards, V. J., & Croft, J. B. (2002). Adverse childhood experiences and personal alcohol abuse as an adult. *Addictive Behaviors, 27*(5), 713–725.

Dube, S. R., Felitti, V. J., Dong, M., Chapman, D. P., Giles, W. H., & Anda, R. F. (2003). Childhood abuse, neglect, and household dysfunction and the risk of illicit drug use: The adverse Childhood Experiences Study. *Pediatrics, 111*(3), 564–572.

Dupré, J. (2007). Fact and value. In H. Kincaid, J. Dupré, & A. Wylie (Eds.), *Value-free science? Ideals and ILLUSIONS* (pp. 27–41). New York: Oxford University Press.

Easterlin, R. A. (1974). Does economic growth improve the human lot? Some empirical evidence. *Nations and Households in Economic Growth, 89,* 89–125.

Einarsdóttir, J. (2012). Happiness in the neonatal intensive care unit: Merits of ethnographic fieldwork. *International Journal of Qualitative Studies on Health and Well-Being, 7,* 1–9.

Ereshefsky, M. (1998). Species pluralism and anti-realism. *Philosophy of Science, 65,* 103–120.

Farrington, D. P. (1995). The development of offending and antisocial behaviour from childhood: Key findings from the Cambridge Study in Delinquent Development. *Journal of Child Psychology and Psychiatry, 36*(6), 929–964.

Fayers, P., & Machin, D. (2013). *Quality of life: The assessment, analysis and interpretation of patient-reported outcomes.* Chichester, UK: John Wiley.

Federal Interagency Forum on Child and Family Statistics. (2012). America's children in brief: Key national indicators of well-being, 2012. Retrieved from http://www.childstats.gov. Accessed 4/13/13

Feinberg, J. (1980). A child's right to an open future. In W. Aiken & H. LaFollette (Eds.), *Whose child? Parental rights, parental authority and state power* (pp. 124–153). Totowa, NJ: Littlefield, Adams.

Feldman, F. (2002). The good life: A defense of attitudinal hedonism. *Philosophy and Phenomenological Research, 65*, 604–628.

Feldman, F. (2004). *Pleasure and the good life*. Oxford: Oxford University Press.

Fine, A. (1998). The viewpoint of no-one in particular. *Proceedings and Addresses of the American Philosophical Association, 72*(2), 7–20.

Fishkin, J. (2009). *When the people speak: Deliberative democracy and public consultation*. Oxford: Oxford University Press.

Fitzpatrick, R., Fletcher, A., Gore, S., Jones, D., Spiegelhalter, D. & Cox, D. (1992). Quality of life measures in health care: I. Applications and issues in assessment. *British Medical Journal, 305*, 1074–1077.

Fletcher, A., Gore, S., Jones, D., Fitzpatrick, R., Spiegelhalter, D. & Cox, D. (1992). Quality of life measures in health care: II. Design, analysis and interpretation. *British Medical Journal, 305*, 1145–1148.

Fletcher, G. (2009). Rejecting well-being invariabilism. *Philosophical Papers, 38*(1), 21–34.

Fletcher, G. (2013). A fresh start for the objective-list theory of well-being. *Utilitas, 25*(2), 206–220.

Fletcher, G. (2016). Objective list theories. In G. Fletcher (Ed.), *The Routledge handbook of philosophy of well-being* (pp. 148–160). New York: Routledge.

Fredrickson, B. L. (2001). The role of positive emotions in positive psychology: The broaden-and-build theory of positive emotions. *American Psychologist, 56*(3), 218–226.

Frey, B. S. (2008). *Happiness: A revolution in economics*. Cambridge, MA: MIT Press.

Fujiwara, D., & Campbell, R. (2011). *Valuation techniques for social cost-benefit analysis: Stated preference, revealed preference and subjective well-being approaches: A discussion of the current issues*. London: HM Treasury.

Galobardes, B., Smith, G. D., & Lynch, J. W. (2006). Systematic review of the influence of childhood socioeconomic circumstances on risk for cardiovascular disease in adulthood. *Annals of Epidemiology, 16*(2), 91–104.

Gasper, D. (2010). Understanding the diversity of conceptions of well-being and quality of life. *The Journal of Socio-Economics, 39*(3), 351–360.

Gaus, G. (2011). *The order of public reason: A theory of freedom and morality in a diverse and bounded world*. Cambridge, UK: Cambridge University Press.

Gerritsen, P., & Van Der Ende, P. (1994). The development of a care-giving burden scale. *Age and Ageing, 23*, 483–491.

Gheaus, A. (2014). The "intrinsic goods of childhood" and the just society. In A. Bagattini & C. Macleod (Eds.), *The nature of children's well-being of children: theory and practice* (pp. 35–52). Dordrecht: Springer.

Gheaus, A. (2015). Unfinished adults and defective children: On the nature and value of childhood. *Journal for Ethics and Social Philosophy, 9*(1), 1–21.

Gilbert, D. (2009). *Stumbling on happiness*. New York: Random House.

Glaeser, E. (2011). The moral heart of economics. *The New York Times*, January 25. Retrieved from http://economix.blogs.nytimes.com/2011/01/25/the-moral-heart-of-economics/?scp=3&sq=glaeser&st=cse

Goldstein, J., Freud, A., & Solnit, A. J. (1973). *Beyond the best interests of the child.* New York: Free Press.

Gopnik, A. (2009). *The philosophical baby.* New York: Farrar, Straus and Giroux.

Gopnik, A. (2015). Personal profile. Retrieved from https://www.ted.com/speakers/alison_gopnik

Gould, S. J. (1981). *The mismeasure of man.* New York: W. W. Norton.

Greco, J. (2008). What's wrong with contextualism? *Philosophical Quarterly,* 58(232), 416–436.

Griffin, J. (1986). *Well-being: Its meaning, measurement and moral importance.* Oxford: Clarendon Press.

Griffin, J. (2007). What do happiness studies study? *Journal of Happiness Studies,* 8, 139–148.

Griffiths, P. E. (2002). What is innateness? *The Monist,* 85(1), 70–85.

Griffiths, P. E., & Stotz, K. (2006). Genes in the postgenomic era? *Theoretical Medicine and Bioethics,* 27(6), 499–521.

Groll, D. (2015). Medicine & well-being. In Guy Fletcher (Ed.), *The Routledge handbook of philosophy of well-being* (pp. 504–516). New York: Routledge.

Grossmann, K., & Grossmann, K. E. (2005). The impact of attachment to mother and father and sensitive support of exploration at an early age on children's psychosocial development through young adulthood. In *Encyclopedia on early childhood development* (pp. 1–8). Montreal: Centre of Excellence for Early Childhood Development.

Gruber, J., Mauss, I. B., & Tamir, M. (2011). A dark side of happiness? How, when, and why happiness is not always good. *Perspectives on Psychological Science,* 6(3), 222–233.

Guala, F. (2007). The philosophy of social science: Metaphysical and empirical. *Philosophy Compass,* 2(6), 954–980.

Gul, F., & Pesendorfer, W. (2008). The case for mindless economics. In A. Caplin & A. Shotter (Eds.), *The foundations of positive and normative economics* (pp. 3–42). Oxford: Oxford University Press.

Guyatt, G. H., Feeny, D. H., & Patrick, D. L. (1993). Measuring health-related quality of life. *Annals of Internal Medicine,* 118, 622–629. doi:10.7326/0003-4819-118-8-199304150-00009

Hacking, I. (1995). *Rewriting the soul: Multiple personality and the sciences of memory.* Princeton, NJ: Princeton University Press.

Hausman, D. M. (2015). *Valuing health: Well-being, freedom, and suffering.* New York: Oxford University Press.

Hausman, D. M., & McPherson, M.S. (2006). *Economic analysis, moral philosophy, and public policy.* New York: Cambridge University Press.

Hausman, D. M. & McPherson, M. S. (2009). Preference satisfaction and welfare economics. *Economics and Philosophy, 25*(1), 1–25.

Hawkins, J. S. (2010). The subjective intuition. *Philosophical Studies, 148*(1), 61–68.

Hawthorne, J. (2004). *Knowledge and lotteries.* New York: Oxford University Press.

Hawthorne, S. (2013). *Accidental intolerance: How we stigmatize ADHD and how we can stop.* New York: Oxford University Press.

Haybron, D. M. (2007). Do we know how happy we are? On some limits of affective introspection and recall. *Noûs, 41*(3), 394–428.

Haybron, D. M. (2008). *The pursuit of unhappiness: The elusive psychology of well-being.* New York: Oxford University Press.

Haybron, D. M. (2016). Mental state approaches to wellbeing. In Matthew Adler and Marc Fleurbaey (Eds.), *The oxford handbook of well-being and public policy* (pp. 347–378). New York, Oxford University Press.

Haybron, D. M., & Alexandrova, A. (2013). Paternalism in economics. In C. Coons & M. Weber (Eds.), *Paternalism: Theory and practice* (pp. 157–177). Cambridge, UK: Cambridge University Press.

Haybron, D. M., & Tiberius, V. (2015). Well-being policy: What standard of well-being? *Journal of the American Philosophical Association, 1,* 712–733.

Heathwood, C. (2007). The reduction of sensory pleasure to desire. *Philosophical Studies, 133*(1), 23–44.

Heywood, C. (2010). Centuries of childhood: An anniversary—and an epitaph? *The Journal of the History of Childhood and Youth, 3*(3), 341–365.

Hillis, S. D., Anda, R. F., Felitti, V. J., & Marchbanks, P. A. (2001). Adverse childhood experiences and sexual risk behaviors in women: A retrospective cohort study. *Family Planning Perspectives, 33*(5), 206–211.

HM Treasury. (2003). *The green book: Appraisal and evaluation in central government.* London: Author.

Hood, S. B. (2013). Psychological measurement and methodological realism. *Erkenntnis, 78*(4), 739–761.

Hunt, S. M. (1997). The problem of quality of life. *Quality of Life Research, 6*(3), 205–212.

Hunt, S. M., McKenna, S. P., McEwen, J., Williams, J., & Papp, E. (1981). The Nottingham Health Profile: subjective health status and medical consultations. *Social Science & Medicine. Part A: Medical Psychology & Medical Sociology, 15*(3), 221–229.

Hunt, S. M., McEwen, J., & McKenna, S. P. (1985). Measuring health status: A new tool for clinicians and epidemiologists. *The Journal of the Royal College of General Practitioners, 35,* 185–188.

Huppert, F. A. (2009). Psychological well-being: Evidence regarding its causes and consequences. *Applied Psychology: Health and Well-Being, 1*(2), 137–164.

Huppert, F. A., Baylis, N., & Keverne, B. (2005). *The science of well-being.* New York, Oxford University Press.

Huppert, F. A., & So, T. T. (2013). Flourishing across Europe: Application of a new conceptual framework for defining well-being. *Social Indicators Research, 110*(3), 837–861.

Janack, M. (2002). Dilemmas of objectivity. *Social Epistemology, 16*(3), 267–281.

John, S. (2015). The example of the IPCC does not vindicate the value free ideal: A reply to Gregor Betz. *European Journal for Philosophy of Science, 5*(1), 1–13.

Kagan, S. (1992). The limits of well-being. *Social Philosophy and Policy, 9*, 180–198.

Kahneman, D. (1999). Objective happiness. In D. Kahneman, E. Diener, & N. Schwarz (Eds.), *Well-being: The foundations of hedonic psychology* (pp. 3–25). New York: Russell Sage Foundation.

Kahneman, D., & Deaton, A. (2010). High income improves evaluation of life but not emotional well-being. *Proceedings of the National Academy of Sciences, 107*(38), 16489–16493. doi:10.1073/pnas.1011492107 .

Kahneman, D., & Krueger, A. B. (2006). Developments in the measurement of subjective well-being. *Journal of Economic Perspectives, 20*(1), 3–24.

Kahneman, D., Krueger, A. B., Schkade, D. A., Schwarz, N., & Stone, A. A. (2004a). A survey method for characterizing daily life experience: The day reconstruction method. *Science, 306*(5702), 1776–1780.

Kahneman, D., Krueger, A., Schkade, D., Schwarz, N., & Stone, A. (2004b). Toward national well-being accounts. *American Economic Review, 94*, 429–434.

Kahneman D., & Riis, J. (2005). Living, and thinking about it: Two perspectives on life. In F. A. Huppert, N. Baylis, & B. Keverne (Eds.), *The science of well-being* (pp. 285–304). Oxford: Oxford University Press.

Kahneman, D., Wakker, P. P., & Sarin, R. (1997). Back to Bentham? Explorations of experienced utility. *The Quarterly Journal of Economics, 112*(2), 375–405.

Kaplan, D. (1989). Demonstratives. In J. Almog, J. Perry, H. K. Wettstein, & D. Kaplan (Eds.), *Themes from Kaplan* (pp. 481–563). Oxford: Oxford University Press.

Kauppinen, A. (2015). The narrative calculus. In Mark Timmons (Ed.), *Oxford Studies in Normative Ethics* 5 (pp. 196–220). Oxford: Oxford University Press.

Keller, S. (2004). Welfare and the achievement of goals. *Philosophical Studies, 121*(1), 27–41.

Keller, S. (2009). Welfare as success. *Noûs, 43*(4), 656–683.

Keyes, C. L., Shmotkin, D., & Ryff, C. D. (2002). Optimizing well-being: The empirical encounter of two traditions. *Journal of Personality and Social Psychology, 82*(6), 1007–1022.

Kim-Prieto, C., Diener, E., Tamir, M., Scollon, C. N., & Diener, M. (2005). Integrating the diverse definitions of happiness: A time-sequential framework of subjective well-being. *Journal of Happiness Studies, 6*, 261–300.

Kingma, E. (2014). Naturalism about health and disease: Adding nuance for progress. *Journal of Medicine and Philosophy, 39*(6), 590–608.

Kirchin, S. (Ed.). (2013). *Thick concepts*. Oxford: Oxford University Press.

Kitcher, P. (2011). *Science in a democratic society*. Amherst, MA: Prometheus Books.

Kopinak, J. K. (1999). The use of triangulation in a study of refugee well-being. *Quality and Quantity, 33*(2), 169–183.

Kourany, J. A. (2003). A philosophy of science for the twenty-first century. *Philosophy of Science, 70*(1), 1–14.

Krantz, D. H., Luce, R. D., Suppes, P., & Tversky, A. (1971). *Foundations of measurement*, Vol. 1: *Additive and polynomial representations*. San Diego and London: Academic Press.

Kraut, R. (2007). *What is good and why: The ethics of well-being*. Cambridge, MA: Harvard University Press.

Kristjánsson, K. (2013). *Virtues and vices in positive psychology*. Cambridge, UK: Cambridge University Press.

Kuhn, T. S. (1962). *The structure of scientific revolutions*. Chicago: University of Chicago Press.

Lacey, H. (1999). *Is science value free? Values and scientific understanding*. London: Routledge.

Lacey, H. (2003). The behavioral scientist qua scientist makes value judgments. *Behavior and Philosophy, 31*, 209–223.

Lacey, H. (2005). *Values and objectivity in science: The current controversy about transgenic crops*. Lanham, MD: Lexington Books.

Lacey, H. (2013). Rehabilitating neutrality. *Philosophical Studies, 163*(1), 77–83.

Lancy, D. F. (2014). *The anthropology of childhood: Cherubs, chattel, changelings*. Cambridge, UK: Cambridge University Press.

Land, K. C., Lamb, V. L., & Mustillo, S. K. (2001). Child and youth well-being in the United States, 1975–1998: Some findings from a new index. *Social Indicators Research, 56*(3), 241–318.

Lawn, P. A. (2003). A theoretical foundation to support the Index of Sustainable Economic Welfare (ISEW), Genuine Progress Indicator (GPI), and other related indexes. *Ecological Economics, 44*(1), 105–118.

Layard, R. (2005). *Happiness: Lessons from a new science*. London: Penguin Books.

Layard, R. (2010). Measuring subjective well-being. *Science, 327*(5965), 534–535.

Lazarus, R. S. (2003). Does the positive psychology movement have legs? *Psychological Inquiry, 14*(2), 93–109.

Legatum Institute. (2015). *The Legatum Prosperity Index 2010 methodology*. Retrieved from http://www.prosperity.com/#!/methodology

Lepore, E., & Cappelen, H. (2005). *Insensitive semantics: A defense of semantic minimalism and speech act pluralism*. Oxford: Blackwell.

Leventhal, T., & Brooks-Gunn, J. (2003). Moving to opportunity: An experimental study of neighborhood effects on mental health. *American Journal of Public Health, 93*(9), 1576–1582.

Liao, S. M. (2006). The right of children to be loved. *Journal of Political Philosophy, 14*(4), 420–440.

Lindeman, M., & Verkasalo, M. (2005). Measuring values with the short Schwartz's value survey. *Journal of Personality Assessment, 85*(2), 170–178.

Lips, P., Cooper, C., Agnusdei, D., Caulin, F., Egger, P., Johnell, O., . . . Wiklund, I. (1997). Quality of life as outcome in the treatment of osteoporosis: The development of a questionnaire for quality of life by the European Foundation for Osteoporosis. *Osteoporos International, 7*, 36–8.

Lips, P., Cooper, C., Agnusdei, D., Caulin, F., Egger, P., Johnell, O., . . . Wiklund, I. (1999). Quality of life in patients with vertebral fractures: Validation of the Quality of Life Questionnaire of the European Foundation for Osteoporosis (QUALEFFO). *Osteoporos International, 10*, 150–160.

Little, D. (1991). *Varieties of social explanation: An introduction to the philosophy of social science.* Boulder, CO: Westview Press.

Loewenstein, G. (2009). That which makes life worthwhile. In *Measuring the subjective well-being of nations: National accounts of time use and well-being* (pp. 87–106). Chicago: University of Chicago Press.

Longino, H. E. (1990). *Science as social knowledge: Values and objectivity in scientific inquiry.* Princeton, NJ: Princeton University Press.

Longino, H. E. (2008). Values, heuristics, and the politics of knowledge. In M. Carrier, D. Howard, & J. Kourany (Eds.), *The challenge of the social and the pressure of practice: science and values revisited* (pp. 68–85). Pittsburgh: University of Pittsburgh Press.

Longino, H. E. (2013). *Studying human behavior: How scientists investigate aggression and sexuality.* Chicago: University of Chicago Press.

Lucas, R. E. (2007). Adaptation and the set-point model of subjective well-being: Does happiness change after major life events? *Current Directions in Psychological Science, 16*(2), 75–79.

Lucas, R. E. (2013). Does life seem better on a sunny day? Examining the association between daily weather conditions and life satisfaction judgments. *Journal of Personality and Social Psychology, 104*, 872–884.

Lucas, R. E., Diener, E., & Suh, E. (1996). Discriminant validity of well-being measures. *Journal of Personality and Social Psychology, 71*(3), 616–628.

Lucas, R. E., Oishi, S., & Diener, E. (2016). What we know about context effects in self-report surveys of well-being: Comment on Deaton and Stone. *Oxford Economic Papers, 68*(4), 871–876.

Lyubomirsky, S., & Lepper, H. S. (1999). A measure of subjective happiness: Preliminary reliability and construct validation. *Social Indicators Research, 46*(2), 137–155.

Macleod, C. (2010). Primary goods, capabilities and children. In H. Brighouse & I. Robeyns (Eds.), *Measuring justice: Primary goods and capabilities* (pp. 174–192). Cambridge, UK: Cambridge University Press.

Matthews, G. (2010). The philosophy of childhood. In Edward N. Zalta (Ed.), *Stanford encyclopedia of philosophy* (Winter 2010 ed.). Retrieved from http://plato.stanford.edu/archives/win2010/entries/childhood/

McAuley, C., & Rose, W. (2011). *Child well-being: Understanding children's lives.* London: Jessica Kingsley.

McClimans, L. (2010). A theoretical framework for patient-reported outcome measures. *Theoretical Medicine and Bioethics, 31*(3), 225–240.

McClimans, L. (2017). Measurement in medicine and beyond: Quality of life, blood pressure and time. In A. Nordman (Ed.), *Reasoning in medicine* (pp. 133–146). New York and London: Routledge.

McClimans, L. M., & Browne, J. (2011). Choosing a patient-reported outcome measure. *Theoretical Medicine and Bioethics, 32*(1), 47–60.

McGillaray, M., & Clarke, M. (Eds.). (2006). *Understanding human well-being.* Tokyo: United Nations University Press.

McKee-Ryan, F., Song, Z., Wanberg, C. R., & Kinicki, A. J. (2005). Psychological and physical well-being during unemployment: A meta-analytic study. *Journal of Applied Psychology, 90*(1), 53–76.

Michell, J. (1999). *Measurement in psychology.* Cambridge, UK: Cambridge University Press.

Mill, J.S. (1882). *A system of logic, ratiocinative and inductive.* Project Gutenberg.

Mokkink, L. B., Terwee, C. B., Patrick, D. L., Alonso, J., Stratford, P. W., Knol, D. L., . . . de Vet, H. C. (2010). The COSMIN checklist for assessing the methodological quality of studies on measurement properties of health status measurement instruments: an international Delphi study. *Quality of Life Research, 19*(4), 539–549.

Morrow, V., & Mayall, B. (2009). What is wrong with children's well-being in the UK? Questions of meaning and measurement 1. *Journal of Social Welfare & Family Law, 31*(3), 217–229.

Myers, C. D., & Mendelberg, T. (2013). Political deliberation. In L. Huddy, D. O. Sears & J. S. Levy (Eds.), *The Oxford handbook of political psychology* (2nd ed.) (pp. 699–736). New York: Oxford University Press.

Nagel, E. (1961). *The structure of science: Problems in the logic of scientific explanation.* New York: Harcourt, Brace & World.

National Institute of Health and Clinical Excellence. (2013). *Guide to the methods of technology appraisal.* London: Author. https://www.nice.org.uk/process/pmg9/chapter/foreword

Nelson, C. A., Zeanah, C. H., Fox, N. A., Marshall, P. J., Smyke, A. T., & Guthrie, D. (2007). Cognitive recovery in socially deprived young children: The Bucharest Early Intervention Project. *Science, 318*(5858), 1937–1940.

Norcross, A. (2005). Harming in context. *Philosophical Studies, 123,* 149–173.

Nord, E. (1999). *Cost-Value Analysis in Health Care: Making Sense of QALYs.* New York: Cambridge University Press.

Northcott, R. (2015) Harm and causation. *Utilitas, 27*(2), 147–164.

Nussbaum, M. C. (2000). *Women and human development: The capabilities approach.* Cambridge, UK: Cambridge University Press.

Nussbaum, M. C. (2001). Political objectivity. *New Literary History, 32*(4), 883–906.

Nussbaum, M., & Sen, A. (1993). *The quality of life*. Oxford: Oxford University Press.

Office of National Statistics. (2012). Proposed domain and headline indicators for measuring national well-being. Retrieved from http://www.ons.gov.uk/ons/about-ons/get-involved/consultations/archived-consultations/2012/measuring-national-well-being-domains/index.html

Office of National Statistics. (2013). Personal well-being across the UK 2012–2013. Retrieved from http://www.ons.gov.uk/ons/dcp171778_328486.pdf

Oishi, S., Schimmack, U., & Colcombe, S. J. (2003). The contextual and systematic nature of life satisfaction judgments. *Journal of Experimental Social Psychology, 39*, 232–247.

Organisation for Economic Co-operation and Development. (2013). OECD guidelines on measuring subjective well-being. Retrieved from http://dx.doi.org/10.1787/9789264191655-en

Parfit, D. (1984). *Reasons and persons*. Oxford: Oxford University Press.

Perry, J. (1998). Indexicals, contexts and unarticulated constituents. In Atocha Aliseda-Llera, Rob J. Van Glabbeek, & Dag Westerståhl (Eds.), *Proceedings of the 1995 CSLI-Armsterdam Logic, Language and Computation Conference* (pp.1–11). Stanford, CA.

Pinnick, C. L., Koertge, N., & Almeder, R. F. (Eds.). (2003). *Scrutinizing feminist epistemology: An examination of gender in science*. New Brunswick, NJ: Rutgers University Press.

Porter, T. (1995). *Trust in numbers: The pursuit of objectivity in science and public life*. Princeton, NJ: Princeton University Press.

Putnam, H. (2002). *The collapse of the fact/value dichotomy and other essays*. Cambridge, MA: Harvard University Press.

Qvortrup, J. (1999). The meaning of child's standard of living. In A. B. Andrews & N. H. Kaufman (Eds.), *Implementing the U.N. Convention on the Rights of the Child: A standard of living adequate for development*. Westport, CT: Praeger.

Raghavan, R., & Alexandrova, A. (2015). Toward a theory of child well-being. *Social Indicators Research, 121*(3), 887–902.

Raibley, J. (2010). Well-being and the priority of values. *Social Theory and Practice, 36*(4), 593–620.

Railton, P. (1986). Facts and values. *Philosophical Topics, 14*(2), 5–31.

Rawls, J. (1993). *Political liberalism*. New York: Columbia University Press.

Reiss, J. (2010). *Error in economics: Towards a more evidence-based methodology*. London: Routledge.

Rigby, M. J., Köhler, L. I., Blair, M. E., & Metchler, R. (2003). Child Health Indicators for Europe: A priority for a caring society. *The European Journal of Public Health, 13*, 38–46.

Risjord, M. (2014). *Philosophy of social science: A contemporary introduction*. New York: Routledge.

Roberts, D. (2013). Thick concepts. *Philosophy Compass, 8*, 677–688. doi:10.1111/ phc3.12055

Robeyns, Ingrid. (2011). The capability approach. In Edward N. Zalta (Ed.), *The Stanford encyclopedia of philosophy* (Summer 2011 ed.). Retrieved from http:// plato.stanford.edu/archives/sum2011/entries/capability-approach/

Rodogno, R. (2014). Well-being, science, and philosophy. In J. H. Søraker, J.-W. van der Rijt, J. de Boer, P.-H. Wong, & P. Brey (Eds.), *Well-being in contemporary society* (pp. 39–58). Dordrecht: Springer.

Root, M. (2007). Social problems. In H. Kincaid, J. Dupré, & A. Wylie (Eds.), *Value-free science? Ideals and illusions* (pp. 42–57). New York: Oxford University Press.

Rosati, C. (1995). Persons, perspectives, and full information accounts of the good. *Ethics, 105*, 296–325.

Rose, N. (1990). *Governing the soul: the shaping of the private self.* New York: Routledge.

Rose, N. (1998). *Inventing our selves: Psychology, power, and personhood.* New York: Cambridge University Press.

Rosenbaum, Sara. (1992). Child health and poor children. *American Behavioral Scientist, 35*(3), 275–289.

Roth, A. (2012). Ethical progress as problem-resolving. *Journal of Political Philosophy, 20*(4), 384–406.

Roth, A. (2013). A procedural, pragmatist account of ethical objectivity. *Kennedy Institute of Ethics Journal, 23*(2), 169–200.

Ryan, R. M., & Deci, E. L. (2001). On happiness and human potentials: A review of research on hedonic and eudaimonic well-being. *Annual Review of Psychology, 52*(1), 141–166.

Ryff, C. (1989). Happiness is everything, or is it? Explorations on the meaning of psychological wellbeing. *Journal of Personality and Social Psychology, 57,* 1069–1081.

Rysiew, P. (2007). Epistemic contextualism. In Edward N. Zalta (Ed.), *Stanford encyclopedia of philosophy.* Retrieved from http://plato.stanford.edu/entries/ contextualism-epistemology/

Sarch, A. F. (2012). Multi-component theories of well-being and their structure. *Pacific Philosophical Quarterly, 93*(4), 439–471.

Scanlon, T. M. (1998). *What we owe to each other.* Cambridge, MA: Harvard University Press.

Schapiro, T. (1999). What is a child? *Ethics, 109,* 715–738.

Schneider, L., & Schimmack, U. (2009). Self-informant agreement in well-being ratings: A meta-analysis. *Social Indicators Research, 94,* 363–376.

Schore, A. N. (1994). *Affect regulation and the origin of the self: The neurobiology of emotional development.* Hove: Psychology Press.

Schwarz, N., & Strack, F. (1991). Evaluating one's life: A judgment model of subjective well-being. In F. Strack, M. Argyle, & N. Schwarz (Eds.), *Subjective well-being* (pp. 27–47). Elmsford, NY: Pergamon Press.

Schwarz, N., & Strack, F. (1999). Reports of subjective well-being: Judgmental processes and their methodological implications. In D. Kahneman, E. Diener, & N. Schwarz (Eds.), *Well-being: The foundations of hedonic psychology* (pp. 61–84). New York: Russell Sage Foundation.

Schwitzgebel, E. (2008). The unreliability of naive introspection. *Philosophical Review, 117*(2), 245–273.

Seaberg, J. R. (1990). Child well-being: A feasible concept? *Social Work, 35*(3), 267–272.

Self, A., Thomas, J., & Randall, C. (2012). Measuring national well-being: life in the UK, 2012. London: Office of National Statistics.

Seligman, M. E. P. (2004). *Authentic happiness: Using the new positive psychology to realize your potential for lasting fulfillment.* New York: Simon & Schuster.

Seligman, M. E., & Csikszentmihalyi, M. (2000). Special issue on happiness, excellence, and optimal human functioning. *American Psychologist, 55*(1), 5–183.

Shonkoff, J. P., & Phillips, D. A. (2000). *From neurons to neighborhoods: The science of early childhood development.* Washington, DC: National Academy Press.

Simms, Leonard J. (2008). Classical and modern methods of psychological scale construction. *Social and Personality Psychology Compass, 2*(1), 414–433.

Sireci, S. G. (1998). The construct of content validity. *Social Indicators Research, 45*(1–3), 83–117.

Sirgy, M. J. (2002). *The psychology of quality of life: Hedonic well-being, life satisfaction, and eudaimonia.* Dordrecht: Kluwer Academic.

Sirgy, M. J. (2012). *The psychology of quality of life: Hedonic well-being, life satisfaction, and Eudaimonia,* Vol. 50. Dordrecht: Springer.

Skelton, A. (2014). Utilitarianism, welfare, children. In A. Bagattini & C. Macleod (Eds.), *The nature of children's well-being: Theory and practice* (pp. 85–103). Dordrecht: Springer.

Skelton, A. (in press). Two conceptions of children's welfare. *Journal of Practical Ethics.*

Sobel, D. (2009). Subjectivism and idealization. *Ethics, 119*(2), 336–352.

Sobel, D. (2011). The limits of the explanatory power of developmentalism. *Journal of Moral Philosophy, 7*(4), 517–527.

Sonenscher, M (2009). The moment of social science: The decade philosophique and late eighteenth century French thought. *Modern Intellectual History, 6,* 121–146. doi:10.1017/S1479244308001960.

Sroufe, L. A., & Rutter, M. (1984). The domain of developmental psychopathology. *Child Development, 55,* 17–29.

Stanley, J. (2004). On the linguistic basis for contextualism. *Philosophical Studies, 119*(1–2), 119–146.

Stanley, J. (2005). *Knowledge and practical interests.* New York and Oxford: Oxford University Press.

Stanley, J. (2007). Précis of knowledge and practical interests and replies to critics. *Philosophy and Phenomenological Research, 75*(1), 168–172.

Starfield, B. (1982). Family income, ill health, and medical care of U.S. children. *Journal of Public Health Policy, 3*(3), 244–259.

Starfield, B. (1992). Effects of poverty on health status. *Bulletin of the New York Academy of Medicine, 68*(1), 17–24.

Steel, D. (2010). Naturalism and the Enlightenment ideal: Rethinking a central debate in the philosophy of social science. In P. D. Magnus & J. Busch (Eds.), *New waves in philosophy of science* (pp. 226–249). London: Palgrave Macmillan.

Stegenga, J. (2015). Effectiveness of medical interventions. *Studies in History and Philosophy of Science Part C: Studies in History and Philosophy of Biological and Biomedical Sciences, 54*, 34–44.

Stevenson, B., & Wolfers, J. (2008). *Economic growth and subjective well-being: Reassessing the Easterlin paradox* (No. w14282). Cambridge, MA: National Bureau of Economic Research.

Stevenson, E. G., & Worthman, C. M. (2014). Child well-being: Anthropological perspectives. In A. Ben-Arieh, F. Casas, & J. E. Korbin (Eds.), *Handbook of child well-being* (pp. 485–512). Dordrecht: Springer.

Stiglitz, J. E., Sen, A., & Fitoussi, J.-P. (2010). *Report by the Commission on the Measurement of Economic Performance and Social Progress.* Paris: Commission on the Measurement of Economic Performance and Social Progress.

Strack, F., Schwarz, N., Chassein, B., Kern, D., & Wagner, D. (1990). The salience of comparison standards and the activation of social norms: Consequences for judgments of happiness and their communication. *British Journal of Social Psychology, 29*, 303–14.

Strauss, M. E., & Smith, G. T. (2009). Construct validity: Advances in theory and methodology. *Annual Review of Clinical Psychology, 5*, 1–25.

Stull, D. E., Kosloski, K., & Kercher, K. (1994). Caregiver burden and generic well-being: Opposite sides of the same coin? *The Gerontologist, 34*(1), 88–94.

Sumner, L. W. (1996). *Welfare, happiness, and ethics.* Oxford: Oxford University Press.

Suppes, P. (1998). Measurement, theory of. In *The Routledge Encyclopedia of Philosophy.* Taylor and Francis. Retrieved 2 March 2017, from https://www.rep.routledge.com/articles/thematic/measurement-theory-of/v-1. doi:10.4324/9780415249126-Q066-1

Takai, M., Takahashi, M., Iwamitsu, Y., Ando, N., Okazaki, S., Nakajima, K., . . . Miyaoka, H. (2009). The experience of burnout among home caregivers of patients with dementia: relations to depression and quality of life. *Archives of Gerontology and Geriatrics, 49*(1), 1–5.

Tal, E. (2013). Old and new problems in philosophy of measurement. *Philosophy Compass, 8*(12), 1159–1173.

Tal, E. (2016). Measurement in science. In Edward N. Zalta (Ed.), *The Stanford Encyclopedia of Philosophy* (Winter 2016 ed.). Retrieved from https://plato.stanford.edu/archives/win2016/entries/measurement-science/

Taylor, C. (1971). Interpretation and the sciences of man. *The Review of Metaphysics, 25*(1), 3–51.

Taylor, T. E. (2015). The markers of wellbeing: A basis for a theory-neutral approach. *International Journal of Wellbeing, 5*(2), 75–90.

Tiberius, V. (2004). Cultural differences and philosophical accounts of well-being. *Journal of Happiness Studies, 5*(3), 293–314.

Tiberius, V. (2007). Substance and procedure in theories of prudential value. *Australasian Journal of Philosophy, 85*(3), 373–391.

Tiberius, V. (2008). *The reflective life: Living wisely with our limits.* Oxford: Oxford University Press.

Tiberius, V. (2013). Well-being, wisdom and thick theorizing: On the division of labour between moral philosophy and positive psychology. In S. Kirchin (Ed.), *Thick concepts* (pp. 217–234). Oxford: Oxford University Press.

Tiberius, V., & Plakias, A. (2010). Wellbeing. In J. Doris & Moral Psychology Research Group (Eds.), *Anthology of moral psychology* (pp. 402–432). Oxford: Oxford University Press.

Tierney, J. (2011). A new gauge to see what's beyond happiness. *The New York Times*, May 6. Retrieved from http://www.nytimes.com/2011/05/17/science/17tierney.html?_r=1

Unger, P. (1995). Contextual analysis in ethics. *Philosophy and Phenomenological Research, 58*, 1–26.

UNICEF. (1979–). The state of the world's children. New York: Author.

United Nations Children's Fund. (2012). The state of the world's children 2012: Children in an Urban World. Retrieved from http://www.unicef.org/sowc2012/pdfs/SOWC 2012-Main Report_EN_13Mar2012.pdf

Van Fraassen, B. C. (2008). *Scientific representation: Paradoxes of perspective.* Oxford: Oxford University Press.

Velleman, J. D. (1991). Well-being and time. *Pacific Philosophical Quarterly, 72*(1), 48–77.

Visser-Meily, A., Post, M., Schepers, V., & Linderman, E. (2005). Spouses' quality of life 1 year after stroke: Prediction at the start of clinical rehabilitation. *Cerebrovascular Diseases, 20*, 443–448.

Wachs, T. D. (1999). *Necessary but not sufficient: The respective roles of single and multiple influences on individual development.* Washington, DC: American Psychological Association.

Warren, S. L., Huston, L., Egeland, B., & Sroufe, L. (1997). Child and adolescent anxiety disorders and early attachment. *Journal of the American Academy of Child & Adolescent Psychiatry, 36*(5), 637–644.

Watson, D., Clark, L. A., & Tellegen, A. (1988). Development and validation of brief measures of positive and negative affect: The PANAS scales. *Journal of Personality and Social Psychology, 54*(6), 1063–1070.

Weber, M. (1949). *The methodology of the social sciences.* Translated by E. A. Shils & H. A. Finch. Glencoe, IL: Free Press.

Welsh, M. C., Pennington, B. F., & Groisser, D. B. (1991). A normative-developmental study of executive function: A window on prefrontal function in children. *Developmental Neuropsychology, 7*(2), 131–149.

Wieland, N. (2011). Parental obligation. *Utilitas, 23*(3), 249–267.

Williams, B. (1985). *Ethics and the limits of philosophy.* Cambridge, MA: Harvard University Press.

Williamson, D. F., Thompson, T. J., Anda, R. F., Dietz, W. H., & Felitti, V. (2002). Body weight and obesity in adults and self-reported abuse in childhood. *International Journal of Obesity and Related Metabolic Disorders, 26*(8), 1075–1082.

Wilson, J. (2014). *Mindful America: The mutual transformation of Buddhist meditation and American culture.* New York: Oxford University Press.

Wilson, M. (2006). *Wandering significance: An essay on conceptual behavior.* Oxford: Oxford University Press.

Wilson, M. (2013). Using the concept of a measurement system to characterize measurement models used in psychometrics. *Measurement, 46*(9), 3766–3774.

Wilson, T. D. (2009). *Strangers to ourselves: Discovering the adaptive unconscious.* Cambridge, MA: Harvard University Press.

Woodard, C. (2013). Classifying theories of welfare. *Philosophical Studies, 165*(3), 787–803.

Woodard, C. (2015). Hybrid theories. In G. Fletcher (Ed.), *Routledge companion to philosophy of well-being* (pp. 161–174). New York: Routledge.

World Health Organization. (1948). Constitution. Geneva: Author. Retrieved from http://whqlibdoc.who.int/hist/official_records/constitution.pdf

Wynne, B. (1989). Sheepfarming after Chernobyl: A case study in communicating scientific information. *Environment: Science and Policy for Sustainable Development, 31*(2), 10–39.

Zacks, J. M., & Maley, C. J. (2007). What's hot in psychology? *APS Observer, 20,* 23–26.

INDEX